8th International Symposium on Symmetries in Subatomic Physics (SSP 2022)

EPJ Web of Conferences Volume 282 (2023)

Vienna, Austria
29 August - 2 September 2022

Editors:

Angela Gligorova
Claude Amsler
Martin Simon
Eberhard Widmann

ISBN: 978-1-7138-7251-1

Printed from e-media with permission by:

Curran Associates, Inc.
57 Morehouse Lane
Red Hook, NY 12571

Some format issues inherent in the e-media version may also appear in this print version.

For additional information, please contact EDP Sciences – Web of Conferences at the address below.

EDP Sciences – Web of Conferences
17, Avenue du Hoggar
Parc d'Activité de Courtabœuf
BP 112
F-91944 Les Ulis Cedex A
France

Phone: +33 (0) 1 69 18 75 75
Fax: +33 (0) 1 69 28 84 91

contact-edps@webofconferences.org

Additional copies of this publication are available from:

Curran Associates, Inc.
57 Morehouse Lane
Red Hook, NY 12571 USA
Phone: 845-758-0400
Fax: 845-758-2633
Email: curran@proceedings.com
Web: www.proceedings.com

TABLE OF CONTENTS

8th International Symposium
on Symmetries in Subatomic Physics (SSP2022)

University of Applied Arts, Vienna
29 August – 02 September 2022

Angela Gligorova[1], Claude Amsler[1],
Martin Simon[1] and Eberhard Widmann[1], Editors

[1]Stefan Meyer Institute for Subatomic Physics
Kegelgasse 27, 1030 Vienna, Austria

The International Symposium on Symmetries in Subatomic Physics (SSP) is a conference series initiated in 1994, devoted to recent accomplishments as well as future developments in fundamental symmetries, in atomic, nuclear and particle physics experiments and theory. The previous meetings were held in Taipei, Seattle, Adelaide, Groningen, Victoria and Aachen.

This 8th edition of the Symposium took place in Vienna from 29 August to 02 September 2022, after one-year delay due to the COVID pandemic. Organized by the Stefan Meyer Institute for Subatomic Physics (SMI) of the Austrian Academy of Sciences and hosted by the University of Applied Arts, the conference was held in-person to the great satisfaction of both participants and organizers.

The scientific program consisted of the following topics:

- Symmetries (C, P, T, CP, CPT, lepton flavour violation),
- Symmetries and Interactions (QED, QCD, Electroweak interactions, Lorentz invariance),
- Fundamental interactions,
- CKM matrix (light and heavy flavours),
- Future facilities,
- Application of new technologies.

As proposed by the International Advisory Committee the scientific program this year was more focused on fundamental symmetries and interactions in theory and laboratory experiments, when compared to previous editions, which also included topics such as Dark Matter and Cosmology. 74 participants from 15 countries attended the conference, and in total 51 invited and contributed talks, as well as 17 posters were presented, highlighting scientific achievements worldwide. The posters were refereed, and two prizes sponsored by the Nuclear Physics European Collaboration Committee (NuPECC) were awarded at a ceremony to Alexander Boeschoten from Van Swinderen Institute in Groningen for the poster *Understanding of Systematic Effects in eEDM Searches with diatomic molecules,* and Marlene Tüchler from the Stefan Meyer Institute for the poster *Kaonic Atom X-Ray Spectrocopy with the SIDDHARTA-2 Experiment*.

A public lecture which was supported by NuPECC and took place in the picturesque Festsaal of the Austrian Academy of Sciences was also part of the program. The talk entitled "The underground world of elementary particles" was given in German by Prof. Stefan Paul from the Technical University of Munich. As part of the social program the participants were offered a guided tour at the Museum of Applied Arts in Vienna (MAK).

The conference was also sponsored by FWF (Doktoratskolleg Particles and Interactions (DK-PI)) and L+H Vakuumtechnik. We wish to thank Ms. F. Boes, Ms. C. Dibold, and Ing. D. Pristauz-Telsnigg for assisting us with the organization of the event, the registration of participants, the coffee breaks and the design of the conference poster and booklet. We also thank the MAK museum and the artist Frederick Baker for the wonderful picture on the SSP2022 poster, which is part of his Virtual Reality Experience "Klimt's magic garden".

The articles published in these Proceedings are peer-reviewed but the scientific accuracy and statements made in the papers rest with the sole responsibilities of the authors.

March 2023
Angela Gligorova

on behalf of the Editors

International Advisory Committee

- Klaus Blaum (MPI Heidelberg)
- Marco Pallavicini (Univ. of Genoa)
- Hans Stroeher (Juelich)
- Livia Ludhova (Juelich)
- Naohito Saito (J-Parc)
- Yoshitaka Kuno (Osaka Univ.)
- Klaus Stefan Kirch (PSI)
- Jorg Pretz (Juelich)
- Klaus Jungmann (Univ. of Groningen)
- Yifang Wang (IHEP, Beijing)
- Stephan Paul (TU Munich)

Local Organizing Committee

- Eberhard Widmann (Chair)
- Claude Amsler
- Angela Gligorova (Scientific Secretary)
- Martin Simon
- Carolina Dibold (Secretary)

Statement of Peer review

In submitting conference proceedings to *Web of Conferences*, the editors of the proceedings certify to the Publisher that

1. They adhere to its **Policy on Publishing Integrity** in order to safeguard good scientific practice in publishing.
2. All articles have been subjected to peer review administered by the proceedings editors.
3. Reviews have been conducted by expert referees, who have been requested to provide unbiased and constructive comments aimed, whenever possible, at improving the work.
4. Proceedings editors have taken all reasonable steps to ensure the quality of the materials they publish and their decision to accept or reject a paper for publication has been based only on the merits of the work and the relevance to the journal.

Title, date and place of the conference

8th International Symposium on Symmetries in Subatomic Physics
29 August - 02 September 2022
Vienna, Austria

Proceedings editor(s):

1. Angela Gligorova
2. Claude Amsler
3. Martin Simon
4. Eberhard Widmann

Date and editor's signature

27/03/2023

17, avenue du Hoggar - PA de Courtabœuf – BP 112 - 91944 Les Ulis Cedex A (France)
Tél. : 33 (0)1 69 18 75 75 - Fax : 33(0)1 69 07 45 17 – www.edpsciences.org

EPJ Web of Conferences **282**, 01001 (2023) https://doi.org/10.1051/epjconf/202328201001
SSP 2022

Lepton Flavor Universality at Belle and Belle II

Seema Choudhury[1,*] on behalf of Belle and Belle II collaborations

[1]Iowa State University and High Energy Accelerator Research Organization

Abstract. We report the test of lepton flavor universality at Belle and Belle II experiments considering $b \to c$ and $b \to s$ transitions. We use 711 fb^{-1} and 189 fb^{-1} of data collected at $\Upsilon(4S)$ resonance from electron-positron collisions for Belle and Belle II, respectively. For the $b \to c$ transition, we present the measurement of lepton-flavor-universality in $B^0 \to D^{*-}\ell^+\nu_\ell$ decay, exclusive measurement of $\mathcal{R}(D)$ & $\mathcal{R}(D^*)$ in Belle data, and inclusive $R(X_{e/\mu})$ in Belle II. For the $b \to s$ transitions, we provide measurements of R_K and R_{K^*} using Belle data, measurement of the branching fraction for the decay $B \to K^*\ell\ell$, and test of lepton-flavor-universality in $B \to J/\psi K$ using Belle II data. Additionally, we report the lepton-flavor-universality in $\Omega_c^0 \to \Omega^-\ell^+\nu_\ell$ decay using 922 fb^{-1} of data collected by the Belle detector.

1 Belle and Belle II detectors

KEKB was an asymmetric energy e^+ (3.5 GeV) - e^- (8 GeV) collider with center-of-mass (CM) energy at the $B\overline{B}$ threshold corresponds to 10.58 GeV. Belle [1] detector was situated at the interaction point of the e^+e^-. It has collected 711, 89.5, and 121.1 fb^{-1} data at the CM energies 10.58, 10.52, and 10.86 GeV, respectively from 1999 to 2010. The detector consisted of many sub-detectors to identify different charged and neutral particles. Belle II [2] is the successor of the Belle experiment with e^+ and e^- energies of 4 GeV and 7 GeV respectively, successfully collecting data from 2019. It aims at collecting 50 ab^{-1} of data by 2035 and plans to deliver collision at a peak luminosity of 6.5×10^{35} cm^{-2}s^{-1} [2] (30 times that of KEKB). The increase in peak luminosity is contributed by the reduction in the beam size by 20 times and 1.5 times increase in beam current. The Belle II detector is also upgraded to cope-up with the beam background which increases in proportion to the luminosity. Belle II has collected $\sim$ 424 fb^{-1} data sample to date, having 364, 42, and 19 fb^{-1} at 10.58, 10.52, and 10.75 GeV, respectively. Belle II has a world record instantaneous luminosity of 4.71×10^{34} cm^{-2}s^{-1} on 22nd June 2022.

2 LFU in $b \to c$ transition

The B decays with $b \to c$ tree-level transition are an important probe to test the lepton-flavor-universality (LFU). The ratio of exclusive decays with τ lepton can be used to test the standard model (SM) by measuring $\mathcal{R}(D^{(*)}) = \dfrac{\mathcal{B}(B \to D^{(*)}\tau\nu)}{\mathcal{B}(B \to D^{(*)}\ell\nu)}$, where $\ell = e, \mu$. The combined result from Belle [3–5], BaBar [6], and LHCb [7, 8] has a tension of 3.1σ with

*e-mail: seema.choudhury@kek.jp

 https://doi.org/10.1051/epjconf/202328201001

SM [9]. The LFU test can also be performed by measuring $R(X) = \dfrac{\mathcal{B}(B \to X\tau\nu)}{\mathcal{B}(B \to X\ell\nu)}$, which is the complimentary measurement to $\mathcal{R}(D^{(*)})$ via inclusive reconstruction. The $R(X)$ measurement is more challenging due to the background from a less constrained X system. The LFU can be tested using light leptons by measuring $R(X_{e/\mu})$.

2.1 Measurement of LFU in $B^0 \to D^{*-}\ell\nu_\ell$ at Belle

Belle [10] has performed the LFU measurement in exclusive semileptonic B decay using $B^0 \to D^{*-}(\bar{D}^0(K^+\pi^-)\pi^-)\ell\nu_\ell$ decay with a data sample of 711 fb^{-1}. The analysis is done using an untagged approach, leading to high efficiency with a sizable background. The background from fake D^* is suppressed by constraining D^* momentum in the CM frame to be < 2.45 GeV$/c$. The signal yield is extracted with a 3-dimensional binned maximum likelihood (ML) fit to $\cos\theta_{B,D^*\ell}$ $\left(\frac{2E^*_B E^*_{D^*\ell} - m_B^2 - m^2_{D^*\ell}}{2|p^*_B||p^*_{D^*\ell}|}\right)$, $\Delta M = M(D^* - D^0)$, and lepton momentum (p_ℓ). The results are shown in Fig. 1. The ratio of the branching fractions of modes with electrons

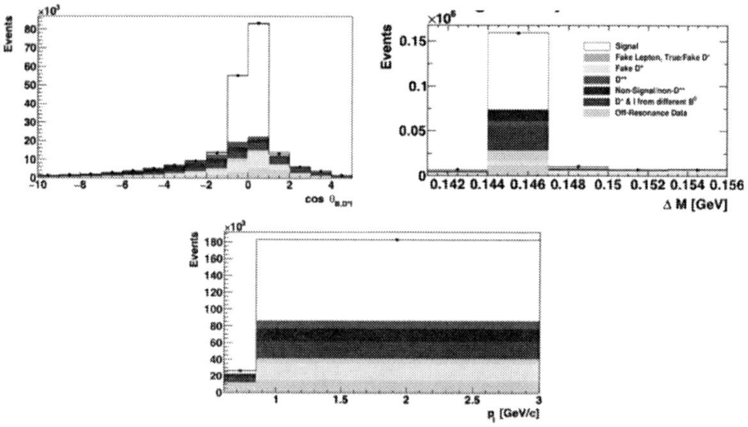

Figure 1. Fits to the $\cos\theta_B$ (top-left), ΔM (top-right), and p_ℓ (bottom) in the μ mode. The points represent the data.

and muons provides a test of the LFU, *i.e.*, $\dfrac{\mathcal{B}(B^0 \to D^{*-}e^+\nu_e)}{\mathcal{B}(B^0 \to D^{*-}\mu^+\nu_\mu)} = 1.01 \pm 0.01(\text{stat}) \pm 0.03(\text{sys})$. The result is consistent with unity within the uncertainty.

2.2 Measurement of $\mathcal{R}(D)$ & $\mathcal{R}(D^*)$ at Belle

Belle [5] has measured simultaneously $\mathcal{R}(D)$ and $\mathcal{R}(D^*)$ using the semileptonic tagging method. Here, the tag-side B is reconstructed via $B^{0/\pm} \to D^{(*)}\ell^-\nu$ decay and signal side B from $B^{0/\pm} \to D^{(*)}\tau^-(\ell^-\bar{\nu}\nu)\nu$, the normalization channel is $B^{0/\pm} \to D^{(*)}\ell^-\nu$. The tagging method ensures good signal purity at the cost of lower signal reconstruction efficiency. The signal is extracted with 2-dimensional binned ML fit to the energies of neutral clusters detected in the ECL that are not associated with any reconstructed particles (E_{ECL}) and the background suppression classifier output (O_{cls}). The fit results are shown in Fig. 2. The measured value of $\mathcal{R}(D) = 0.307 \pm 0.037(\text{stat}) \pm 0.016(\text{sys})$ and $\mathcal{R}(D^*) = 0.283 \pm 0.018(\text{stat}) \pm 0.014(\text{sys})$ have deviations of 0.2σ and 1.1σ, respectively. The combined result of $\mathcal{R}(D)$ & $\mathcal{R}(D^*)$ has a deviation of 0.8σ.

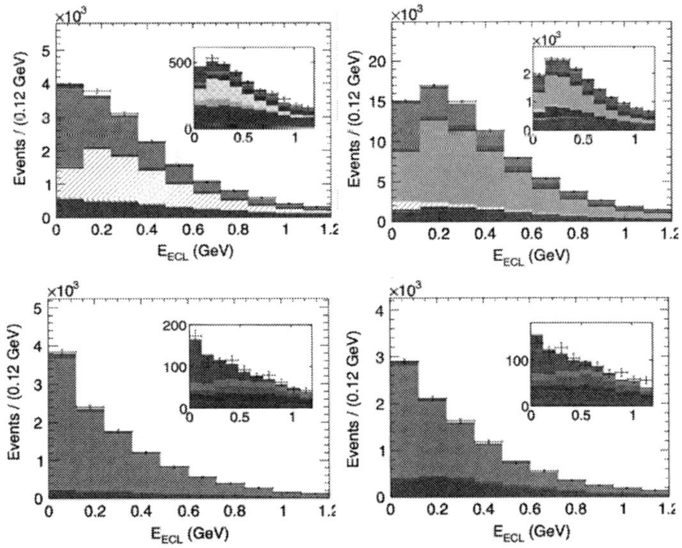

Figure 2. E_{ECL} fit projection and data points with statistical uncertainties in the $D^{+}\ell^{-}$ (top-left), $D^{0}\ell^{-}$ (top-right), $D^{*+}\ell^{-}$ (bottom-left), and $D^{*0}\ell^{-}$ (bottom-right) samples. The signal enhanced region is shown in the inset.

2.3 Measurement of $R(X_{e/\mu})$ at Belle II

Belle II [11] has checked the LFU by inclusive measurement of $R(X_{e/\mu})$ using 189 fb^{-1} of data sample with hadronic-B tagging approach. In this measurement, the signal side B flavor and kinematics in constrained by tagging the other B in its fully hadronic decays, which leads to good signal purity at a cost of lower signal reconstruction efficiency. The X system on the signal side contains a large variety of different charged and neutral final-state particles. As we reconstruct only the lepton in the signal side, the lepton momentum in the B signal rest frame, p^{*}_{ℓ}, is used to extract the signal yield. We required the lepton to have a high probability to be an electron or muon and $p^{*}_{\ell} > 1.3$ GeV/c to suppress backgrounds from hadrons faking as leptons and secondary leptons from $b \to c \to (\ell, s)$ cascades and $B \to X\tau\nu$. The signal yields for $B \to Xe\nu$ and $B \to X\mu\nu$ channels are extracted simultaneously in 10 bins of p^{*}_{ℓ}, with one-dimensional binned ML fit. The fit results are shown in Fig. 3. There are 48034 ± 286, 58569 ± 429 signal events for $B \to Xe\nu$ and $B \to X\mu\nu$ channels, respectively. The $R(X_{e/\mu})$ measured to be $1.033 \pm 0.010(\text{stat}) \pm 0.020(\text{sys})$, for $p^{*}_{\ell} > 1.3$ GeV/c. This is the first inclusive test of (e, μ) lepton flavor universality in semileptonic $B \to X\ell\nu$ decays. The measurement is in agreement with unity within 1.5σ, with a world-leading precision of 2.2%. This measurement paved the path for $R(X) = R(X_{\tau/\ell})$ measurement.

3 LFU in $b \to s$ transition

$b \to s$ is propagated through loop-level transition and is an important probe to test the LFU by measuring R_K and R_{K^*} for $B \to K\ell\ell$ and $B \to K^*\ell\ell$ decays, *i.e.*, $R_{K^{(*)}} = \dfrac{\mathcal{B}(B \to K^{(*)}\mu\mu)}{\mathcal{B}(B \to K^{(*)}ee)}$. According to SM this ratio should be 1 [12], as the coupling of leptons to gauge boson is independent of flavor. LHCb [13] has reported a deviation of 3.1σ in R_{K^+} measurement for $q^2 \in [1.1 - 6.0]$ GeV$^2/c^4$ bin using 9 fb^{-1} data sample, q^2 is invariant mass square of lepton

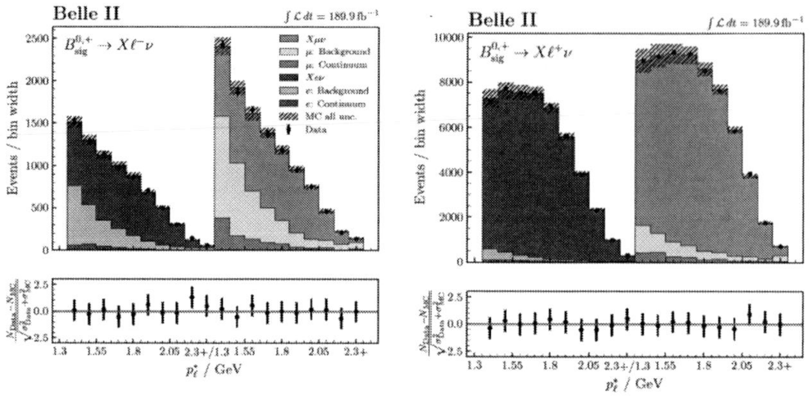

Figure 3. Distributions of Lepton momentum in the $B_{\rm sig}$ frame (p^*_ℓ) in the correct sign (left) and wrong sign (right). The electron and muon templates are fitted simultaneously in $10p^*_\ell$ bins with a width of 100 MeV each covering a p^*_ℓ range from 1.3 GeV to 2.3 GeV.

pair. Similarly, $R_{K^{*+}}$ and $R_{K^0_S}$ measurements from LHCb [14] have 1.4σ and 1.5σ deviations from SM, calculated in 3 fb^{-1} data sample.

3.1 R_{K^*} measurement at Belle

Belle [15] has measured R_{K^*} using a full data sample of 711 fb^{-1}. The decay modes reconstructed are $B^+ \to K^{*+}(K^+\pi^0, K^0_S\pi^+)\ell\ell$ and $B^0 \to K^{*0}(K^+\pi^-, K^0_S\pi^0)\ell\ell$. The kinematic variables which distinguish the signal from the background are beam-energy-constrained mass, $M_{\rm bc} = \sqrt{E^2_{beam}/c^4 - |p_B|^2/c^2}$ and energy difference, $\Delta E = E_B - E_{beam}$. Here, p_B and E_B are the momentum and energy of the B candidate, and E_{beam} is the beam energy. The background coming from the continuum and $B\bar{B}$ are suppressed using Neural Network (NN). The signal yield is obtained by performing a 1-dimensional unbinned extended ML fit in $M_{\rm bc}$, shown in Fig. 4. From the fit, the signal yields are $103^{+13.4}_{-12.7}$ and $139.9^{+16.0}_{-15.4}$ for electron and muon channels, respectively. From the fitted signal yield and signal MC efficiency, the R_{K^*} is calculated for charged B, neutral B, and the combined result from charged and neutral B, as shown in Fig. 5. The results for different bins are consistent with SM expectations within the uncertainty.

3.2 R_K measurement at Belle

R_K tests LFU in $B \to K\ell\ell$ decays. The analysis is performed in 711 fb^{-1} data sample of Belle [16]. The decay modes reconstructed are $B^+ \to K^+\ell\ell$ and $B^0 \to K^0_S\ell\ell$. The background from continuum and $B\bar{B}$ are suppressed using a NN which uses several event shapes, vertex quality, and kinematic variables. The signal yield is extracted by 3-dimensional unbinned extended ML fit in $M_{\rm bc}$, ΔE, and the translated NN output (O'). The fit results are shown in Fig. 6. There are 137 ± 14, 138 ± 15, $27.3^{+6.6}_{-5.8}$, and $21.8^{+7.0}_{-6.1}$ signal events for $B^+ \to K^+\mu\mu$, $B^+ \to K^+ee$, $B^0 \to K^0_S\mu\mu$, and $B^0 \to K^0_S ee$, respectively. The R_{K^+}, R_{K^0}, and R_K are measured in different q^2 bins, as shown in Fig. 7. The results are in agreement with SM expectations within the uncertainty. In addition to this, we have the most precise measurement for R_K in the J/ψ region and is consistent with the unity, *i.e.*, $R_{K^+}(J/\psi) = 0.994 \pm 0.011(\mathrm{stat}) \pm 0.010(\mathrm{sys})$ and $R_{K^0}(J/\psi) = 0.993 \pm 0.015(\mathrm{stat}) \pm 0.010(\mathrm{sys})$.

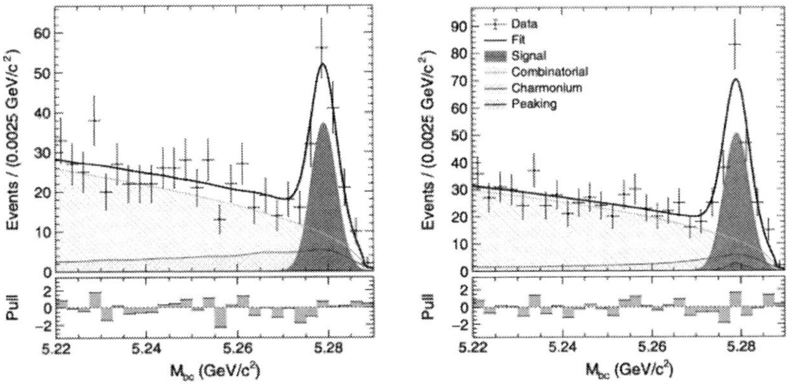

Figure 4. Fit result of $B \to K^*\ell\ell$ in terms of M_{bc} for the electron (left) and muon (right) modes with $q^2 > 0.045$ GeV$^2/c^4$. Combinatorial (dashed blue), signal (red filled), charmonium (dashed green), peaking (purple dotted), and total (solid) fit distributions are superimposed on data (points with error bars).

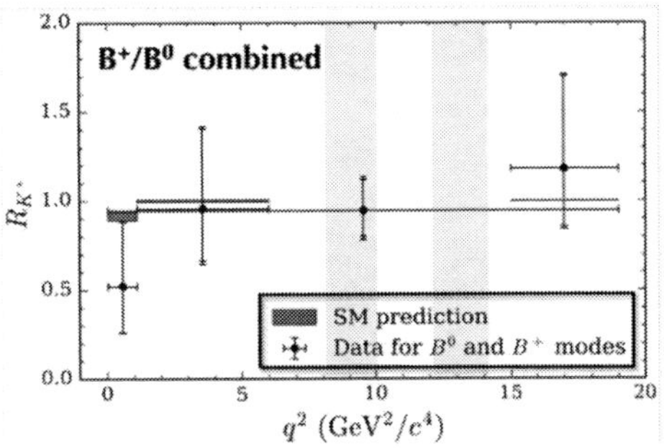

Figure 5. Result for R_{K^*} compared to SM predictions in bins of $q^2 \in [0.045, 1.1]$, $[1.1, 6]$, $[15, 19]$, and $[0.045, 19]$ GeV$^2/c^4$. The shaded bands indicate the charmonium vetoes.

3.3 Measurement of $\mathcal{B}(B \to K^*\ell\ell)$ at Belle II

Belle II [17] has measured $\mathcal{B}(B \to K^*\ell\ell)$ using 190 fb^{-1} of data. For this analysis, the decay modes reconstructed are $B^0 \to K^{*0}(K^+\pi^-)\ell\ell$ and $B^+ \to K^{*+}(K^+\pi^0, K^0_S\pi^+)\ell\ell$. The signal yield is extracted by 2-dimensional un-binned extended ML fit in M_{bc} and ΔE for the events which pass tight requirement on the boosted decision tree, used to fight the background from continuum and $B\bar{B}$. The fit results are shown in Fig. 8. The branching fractions are measured for the entire range of the dilepton mass, excluding the mass region to suppress the $B \to K^*\gamma(\to e^+e^-)$ background and the regions compatible with decays of charmonium resonances to be; $\mathcal{B}(B \to K^*(892)\mu^+\mu^-) = (1.19 \pm 0.31(\mathrm{stat})^{+0.08}_{-0.07}(\mathrm{sys})) \times 10^{-6}$, $\mathcal{B}(B \to K^*(892)e^+e^-) = (1.42 \pm 0.48(\mathrm{stat}) \pm 0.09(\mathrm{sys})) \times 10^{-6}$, and $\mathcal{B}(B \to K^*(892)\ell^+\ell^-) = (1.25 \pm 0.30(\mathrm{stat})^{+0.08}_{-0.07}(\mathrm{sys})) \times 10^{-6}$. The results are compatible with the world averages within the uncertainties. This is a preliminary step towards the LFU test using larger data sample of Belle II.

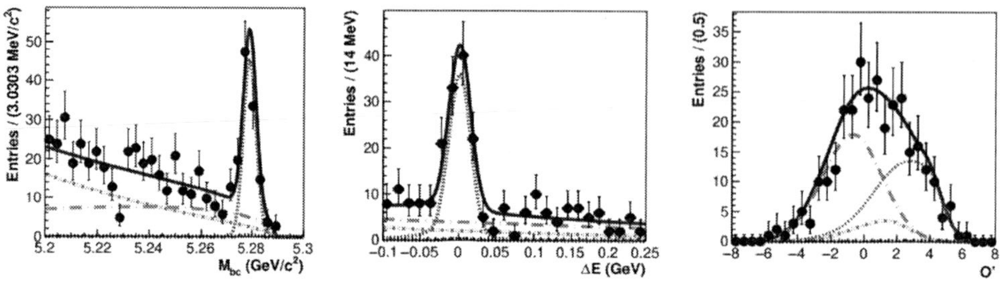

Figure 6. Signal-enhanced M_{bc} (left), ΔE (middle), and $\mathcal{O}'$ (right) projections for $B^+ \to K^+\mu\mu$ decay. Point with error bars are the data, blue solid curves are the fitted result, red dashed curves denote the signal component, cyan long dashed, green dash-dotted, and black dashed curves represent continuum, $B\bar{B}$ background, and $B \to$ charmless decays, respectively.

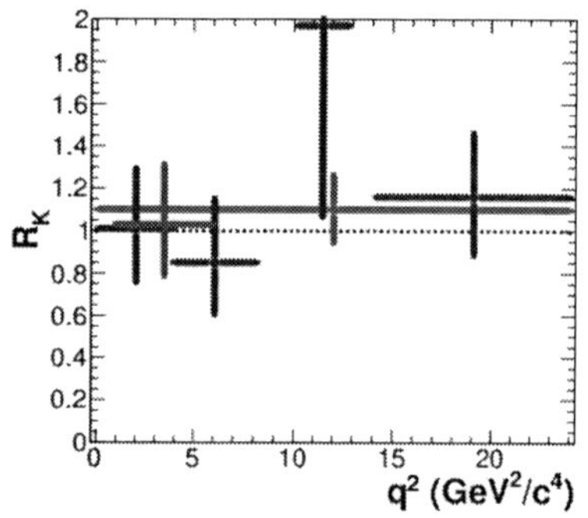

Figure 7. R_K in bins of q^2 for $B \to K\ell\ell$. The red marker represents the bin of $1.0 < q^2 < 6.0$ GeV^2/c^4, and the blue marker are for $0.1 < q^2 < 4.0$, $4.00 < q^2 < 8.12$, $10.2 < q^2 < 12.8$, and $q^2 > 14.18$ GeV^2/c^4 bins. The green marker denotes the whole q^2 region excluding the charmonium resonances.

3.4 Measurement of $R_K(J/\psi)$ at Belle II

Belle II [18] has tested LFU in $B \to J/\psi K$ using 190 fb^{-1} data sample. The decay modes reconstructed are $B^+ \to J/\psi(\ell\ell)K^+$ and $B^0 \to J/\psi(\ell\ell)K^0$. The signal yield is obtained by 2-dimensional un-binned extended ML fit in M_{bc} and ΔE. The fit results are shown in Fig. 9. The signal has a purity of $90-95\%$. The LFU ratio measured from charged and neutral B are $R_{K^+}(J/\psi) = 1.009 \pm 0.022(\mathrm{stat}) \pm 0.008(\mathrm{sys})$ and $R_{K^0}(J/\psi) = 1.042 \pm 0.042(\mathrm{stat}) \pm 0.008(\mathrm{sys})$, respectively. These results are statistically dominated and in agreement with PDG [19].

4 Measurement of $R(\Omega)$ at Belle

LFU in Ω_c^0 is probed for the first time with $\Omega_c^0 \to \Omega^-\ell^+\nu_\ell$ decay using Belle [20] data sample of 922 fb^{-1}. The data of 89.5, 711, and 121.1 fb^{-1} are collected at the CM energies of 10.52, 10.58, and 10.86 GeV, respectively. These decays are reconstructed in the process $e^+e^- \to c\bar{c} \to \Omega_c^0 + anything$. The signal yield is extracted by binned ML fit to the invariant mass ($M_{\Omega\ell}$) spectra. The fit results are shown in Fig. 10. The signal yields for $\Omega_c^0 \to \Omega^-\pi^+$, $\Omega_c^0 \to \Omega^-e^+\nu_e$, and $\Omega_c^0 \to \Omega^-\mu^+\nu_\mu$ are 865.3 ± 35.3, 707.6 ± 37.7, and 367.9 ± 31.4 events,

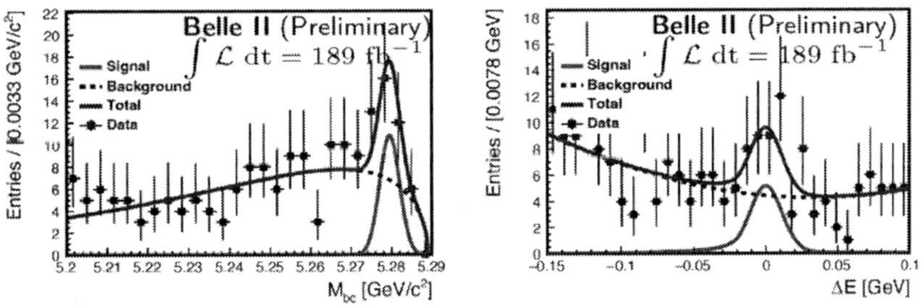

Figure 8. $M_{\rm bc}$ (left) and ΔE (right) projection for $B \to K^*\ell\ell$ decay. Points with error bars are superimposed on the blue (solid) curve showing total fit result, while red (solid) and black (dotted) lines represent the signal and background components.

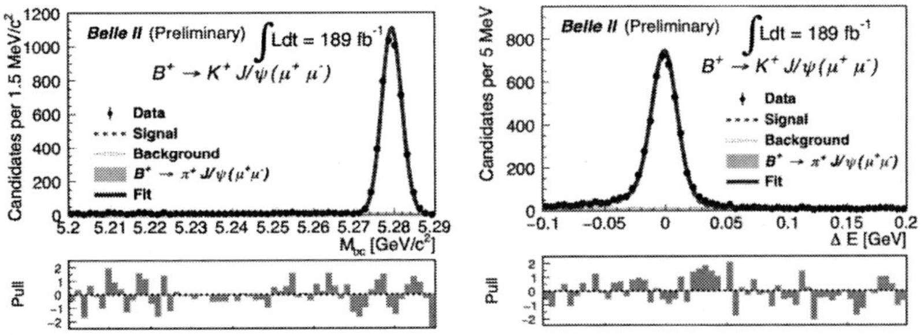

Figure 9. $M_{\rm bc}$ (left) and ΔE (right) distributions with the fit result superimposed. Black dots with error bars denote the data, blue curve denote the total fit, dashed red curves are the signal component, dotted green curves are the background component, and filled cyan regions are the $B^+ \to J/\psi\pi^+$ component.

respectively. The significance of the $\Omega_c^0 \to \Omega^-\ell^+\nu_\ell$ are both larger than 10σ, and $\Omega_c^0 \to \Omega^-\mu^+\nu_\ell$ decay is seen for the first time in Belle, where $\Omega_c^0 \to \Omega^-\pi^+$ decay is used as a normalization channel. The semileptonic decay branching fraction ratio is calculated to be; $R(\Omega) = \dfrac{\mathcal{B}(\Omega_c^0 \to \Omega^-e^+\nu_e)}{\mathcal{B}(\Omega_c^0 \to \Omega^-\mu^+\nu_\mu)} = 1.02 \pm 0.10(\text{stat}) \pm 0.02(\text{sys})$. The $R(\Omega)$ agrees with the expected LFU value 1.03 ± 0.06 [21] within the uncertainty.

5 LFU prospects at Belle II

The $R(X)$ measurement or in general inclusive processes are unique to Belle II. The estimated precision on $R(X)$ for 189 fb^{-1} of data sample is $\sim$ 17%. The uncertainty (statistical + systematic) projections for $\mathcal{R}(D)$, $\mathcal{R}(D^*)$, and $R(X)$ are shown in Fig. 11 [22]. A few ab^{-1} of data from Belle II will be sufficient to track the anomaly on $\mathcal{R}(D) - \mathcal{R}(D^*)$ to be statistical or systematic origin. With the full data sample of Belle II, the total uncertainty for $\mathcal{R}(D) - \mathcal{R}(D^*)$ will be $2-3\%$ for different tagging approaches (semileptonic or hadronic). $R(X)$ will also be of similar precision. Similarly, for $b \to s$ transition, the R_{K^*} projections for 1 ab^{-1}, 5 ab^{-1}, and 50 ab^{-1} are shown in Fig. 11. The R_{K^+} and R_{K^*} statistical sensitivity will be $< 2\%$ for the entire q^2 region and $\sim 3\%$ for $q^2 \in [1-6]\ \mathrm{GeV}^2/c^4$, with full data sample of 50 ab^{-1}. Contrary to LHCb, Belle II can check these observables both for low and high q^2 bins. Belle II will

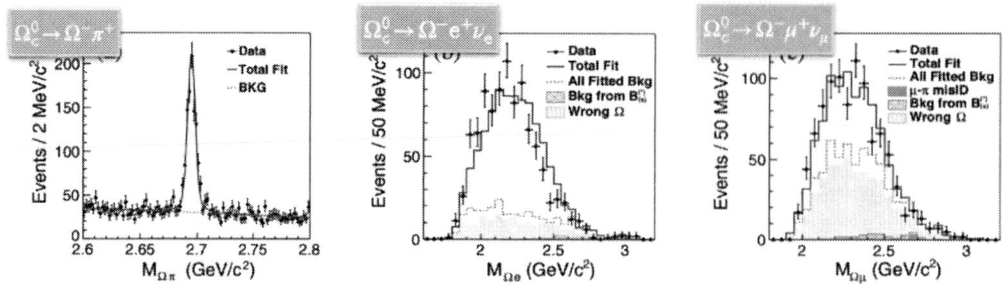

Figure 10. Fit to the $M_{\Omega\pi}$ (left), $M_{\Omega e}$ (middle), $M_{\Omega\mu}$ (right) distributions. The dots with error bars represent the data, the solid lines are the best fits, and the dashed lines are the fitted total background. The blank areas between the red dashed lines and shaded histograms are from backgrounds with mis-selected ℓ^+. The $\mu - \pi$ misID in $M_{\Omega\mu}$ plot is the background from $\Omega_c^0 \to \Omega^-\pi^+ + hadrons$ decays.

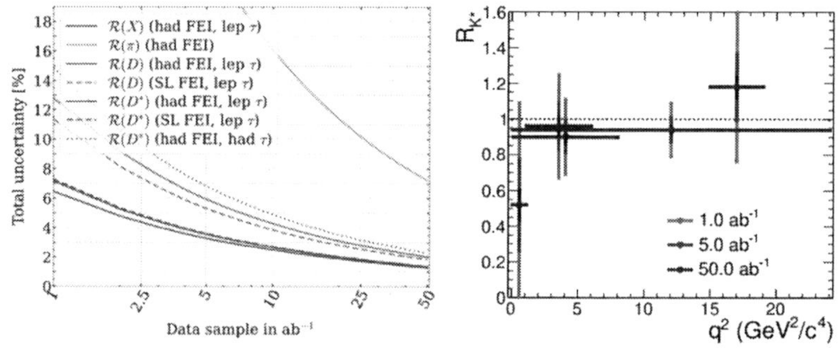

Figure 11. LFU prospects of Belle II for $b \to c$ (left) and $b \to s$ (right) transitions.

provide an independent measurement to confirm the tension observed in $R_K - R_{K^*}$ with few ab^{-1} of data sample.

6 Summary

The flavor physics in e^+e^- collisions offers an extremely rich physics program with many opportunities to probe physics beyond the SM. At Belle II, we can access charged and neutral B with equal efficiency and also have equal sensitivity for electron and muon channels. Belle II can access inclusive decay modes in addition to exclusive decays, and analysis can be performed in tagged or un-tagged approaches. Belle II has collected $\sim$ 424 fb^{-1} of data which is comparable to the size of the BaBar data sample and can be combined with the Belle data sample to increase statistics. So far, Belle or Belle II have not observed any sign of LFU violation and a higher data sample will shed light on LFU anomalies. Some results show the potential of Belle II, and only a part of them are covered here. We will have more results from Belle II in near future.

References

[1] A. Abashian *et al.* (Belle Collaboration), Nucl. Instrum. Methods Phys. Res., Sec. A 479, **117** (2002); also, see the detector section in J. Brodzicka et al., Prog. Theor. Exp. Phys.

2012, 04D001 (2012).
[2] T. Abe *et al.* (Belle II Collaboration), arXiv:1011.0352.
[3] M. Huschle *et al.* (Belle Collaboration), Phys. Rev. D **92**, 072014 (2015).
[4] S. Hirose *et al.* (Belle Collaboration), Phys. Rev. Lett. **118**, 211801 (2017).
[5] G. Caria *et al.* (Belle Collaboration), Phys. Rev. Lett. **124** 161803 (2020).
[6] J. P. Lees *et al.* (BaBar Collaboration), Phys. Rev. D **88**, 072012 (2013).
[7] R. Aaij *et al.* (LHCb Collaboration), Phys. Rev. Lett. **115**, 111803 (2015).
[8] R. Aaij *et al.* (LHCb Collaboration), Phys. Rev. D **97**, 072013 (2018).
[9] Y. Amhis *et al.* (HFLAV Collaboration), arXiv:2206.07501.
[10] E. Waheed *et al.* (Belle Collaboration), Phys. Rev. D **100**, 052007 (2019).
[11] H. Junkerkalefeld *et al.* (Belle II Collaboration), ICHEP 2022.
[12] M. Bordone, G. Isidori, and A. Pattori, Eur. Phys. J. C **76**, 440 (2016).
[13] R, Aaji *et al.* (LHCb Collaboration), Nature Physics **18**, 277 (2022).
[14] A. Aaij *et al.* (LHCb Collaboration), Phys. Rev. Lett. **128**, 191802 (2022).
[15] S. Wehle *et al.* (Belle Collaboration), Phys. Rev. Lett. **126**, 161801 (2021).
[16] S. Choudhury *et al.* (Belle Collaboration), J. High Energy Phys. **03**, 105 (2021)
[17] F. Abudinén *et al.* (Belle II Collaboration), arXiv:2206.05946.
[18] F. Abudinén *et al.* (Belle II Collaboration), arXiv:2207.11275.
[19] P.A. Zyla *et al.* (Particle Data Group), Prog. Theor. Exp. Phys. **2020**, 083C01 (2020).
[20] Y. B. Li *et al.* (Belle Collaboration), Phys. Rev. D **105**, L091101 (2022).
[21] F. Huang, Q. A. Zhang, arXiv:2108.06110.
[22] Snowmass 2021, arXiv: 2207.06307.

Discrete symmetry tests using hyperon-antihyperon pairs

Andrzej Kupsc[1,2,*]

[1]Uppsala University, Sweden
[2]National Centre for Nuclear Research, Poland

Abstract. The BESIII experiment has accumulated millions of spin entangled hyperon–antihyperon pairs from the decays of the J/ψ and $\psi(2S)$ mesons produced in electron–positron collisions. Using multi-dimensional methods to analyze the angular distributions in the sequential decays the hyperon and antihyperon decay amplitudes are studied and tests of the combined charge-conjugation–parity (CP) symmetry are performed. Implications of the recent results using J/ψ decays into $\Lambda\bar{\Lambda}$ and $\Xi^-\bar{\Xi}^+$ are presented. For the cascade decay chain, the exclusive measurement allows for three independent CP tests and the first direct determination of the weak phases.

1 Introduction

The standard model of elementary particles explains most of the subatomic phenomena but it leaves out questions like the unification with the gravitational forces or the origin of the observed dominance of the matter over the antimatter in the universe. Any candidate for the underlying fundamental theory has to address these questions. A clue how to construct such theory might be found by scrutinizing the cornerstone symmetries of the standard model. In addition to the *natural* symmetries like space-time translations and space rotations, which lead to the four-momentum and angular momentum conservation, there are discrete symmetries related to parity (P), charge conjugation (C) and time reversal (T) transformations. Parity is the spatial inversion which changes directions of all particles to the opposite. Charge conjugation is the swap between all particles and antiparticles in a system. Separately the C and P symmetries are maximally violated in the weak interactions but their combination — the CP symmetry — is nearly preserved and a small observed violation is due to quantum interference effects only. The quantum interference can also enhance a signal of new interactions. For example, having the amplitude of a known process $M = |M|\exp(i\delta)$ and a new process amplitude $m = |m|\exp(i\xi)$ the probability of the combined process is $|M+m|^2 = |M|^2 + 2|M||m|\cos(\delta+\xi) + |m|^2$. If $|m|^2 \ll |M|^2$ the contribution of the new process is dominated by the interference term that is linear in $|m|$.

Here I will discuss direct CP violation that requires amplitudes for a process and the CP-transformed process to have different complex phases. For example if the amplitude for a particle decay is $\mathcal{A} = |\mathcal{A}|\exp(i\xi + i\delta)$ then the amplitude for the CP-transformed process is $\bar{\mathcal{A}} = |\mathcal{A}|\exp(-i\xi + i\delta)$, where ξ is the CP-odd phase that changes sign after the CP transformation. In order to observe an effect due to the CP-odd phases an interference with other amplitudes contributing to the process is needed. Since the CP-odd phases are small for the

*e-mail: Andrzej.Kupsc@physics.uu.se

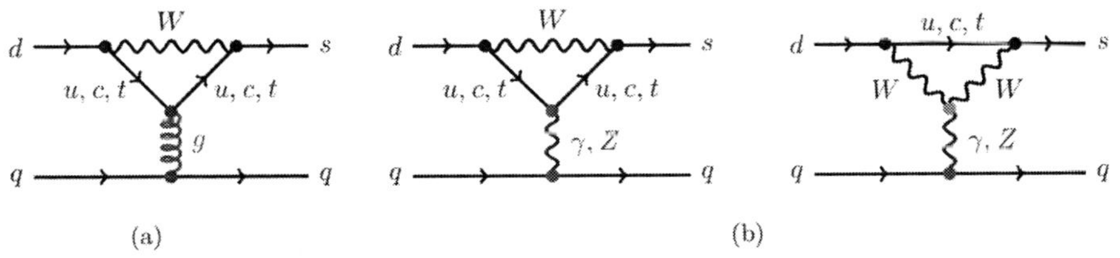

Figure 1. Quark level diagrams for the weak processes involving kaons or hyperons that contribute to the CP-odd phases in the standard model: (a) QCD-penguin operators and (b) electroweak penguin operators. This figure was created using a modified script from Ref. [7].

weak transitions of strange quarks a significant contribution might be due to a new physics (NP) effect. One process where the direct CP-violation was observed is the combined weak and strong decay of K^0 meson into a pion pair $K^0 \to \pi^+\pi^-$. There are two final states that have distinct properties with respect to the strong interactions[1]. The initial weak transitions can proceed to one of the two states and are represented by the amplitudes $\mathcal{A}_0$ and $\mathcal{A}_2$ where $|\mathcal{A}_2/\mathcal{A}_0| \approx 0.05$. The strong interaction of the pions is described by two strong phases δ_0 and δ_2. The complete amplitude for the $K^0 \to \pi^+\pi^-$ process can then be written as $\mathcal{A} = |\mathcal{A}_0|\exp(i\xi_0 + i\delta_0) + |\mathcal{A}_2|\exp(i\xi_2 + i\delta_2)$. The amplitude of the CP-transformed process $\bar{K}^0 \to \pi^-\pi^+$, $\bar{\mathcal{A}}$, differs only by the signs of the CP-odd phases ξ_0 and ξ_2. The decay rates $|\mathcal{A}|^2$ and $|\bar{\mathcal{A}}|^2$ can be different if CP is violated. The direct CP violation observable is expressed as [1, 2]:

$$\mathrm{Re}(\epsilon') := \frac{1}{2}\frac{|\mathcal{A}|^2 - |\bar{\mathcal{A}}|^2}{|\mathcal{A}|^2 + |\bar{\mathcal{A}}|^2} \approx (\xi_0 - \xi_2)\sin(\delta_0 - \delta_2)\left|\frac{\mathcal{A}_2}{\mathcal{A}_0}\right|, \tag{1}$$

where the measured value of $\mathrm{Re}(\epsilon')$ is $(3.7 \pm 0.5) \times 10^{-6}$ [3] corresponding to the weak-phase difference $\xi_0 - \xi_2 \approx 10^{-4}$ rad. In the standard model the contributions to the CP-odd phases $\xi_{0,2}$ are given by the loops that involve all three quark generations as shown in the diagrams in Fig. 1. An order of magnitude estimate for these contributions is determined from the Cabibbo–Kobayashi–Maskawa matrix [4, 5] and can be expressed by the product of the Wolfenstein parameters [6] as $\lambda^4 A^2 \eta \approx -6 \times 10^{-4}$ [3]. Recent standard model predictions for the CP-violation in kaon decays that include hadronic effects are given in the framework of the low energy effective field theory [7] and the lattice quantum chromodynamics [8, 9]. The agreement with the $\mathrm{Re}(\epsilon')$ parameter measurement is satisfactory but due to large uncertainties there is a room for a new physics contribution.

2 Baryon two-body weak decays

Similar mechanism of the direct CP violation is possible in the main decay modes of the baryons with one or more strange quarks — hyperons. The spin of a baryon provides an additional degree of freedom that can be used for more sensitive CP tests. A quantum state of a spin-1/2 baryon B can be described by Pauli matrices:

$$|B\rangle = \sigma_0 + \mathbf{P}_B \cdot \boldsymbol{\sigma}\,, \tag{2}$$

where σ_0 is the 2×2 unit matrix and $\boldsymbol{\sigma} := (\sigma_1, \sigma_2, \sigma_3)$ represents spin x, y, z projections in the baryon rest frame. The polarization vector $\mathbf{P}_B$ describes the preferred spin direction

[1]They have different values of the isospin quantum number $I = 0$ and $I = 2$ for the two-pion system.

for an ensemble of the B-baryons. Consider the $B \to b\pi$ transition between two spin-1/2 baryons with positive internal parities where a negative parity pion is emitted. Examples of such processes are the decays $\Xi^- \to \Lambda\pi^-$ or $\Lambda \to p\pi^-$. The baryon b can have the same or opposite direction of spin as the baryon B and the angular momentum conservation requires s- or p-wave for the b–π system. The parity of the final state is negative for the s-wave and positive for the p-wave. Both are allowed since parity is not conserved in weak decays. The decay amplitude is given by the transition operator

$$\mathcal{A}(B \to b\pi) \propto \mathcal{S}\sigma_0 + \mathcal{P}\sigma \cdot \hat{\mathbf{n}} , \tag{3}$$

where $\hat{\mathbf{n}}$ is the direction of the b-baryon in the B-baryon rest frame. The complex parameters $\mathcal{S}$ and $\mathcal{P}$ can be represented as $\mathcal{S} = |\mathcal{S}|\exp(i\xi_S + i\delta_S)$ and $\mathcal{P} = |\mathcal{P}|\exp(i\xi_P + i\delta_P)$. The strong interaction of the b–π system is given by the δ_P and δ_S phases and the CP-odd phases are ξ_P and ξ_S. The amplitude ratios $|\mathcal{P}/\mathcal{S}|$ for $\Lambda \to p\pi^-$ and $\Xi^- \to \Lambda\pi^-$ are 0.442(4) and 0.188(2), respectively [10]. Since only the angular distributions will be discussed the probability of the decay $\Gamma \propto |\mathcal{S}|^2 + |\mathcal{P}|^2$ is set to $|\mathcal{S}|^2 + |\mathcal{P}|^2 = 1$. The real and imaginary parts of the amplitude interference term are

$$\alpha = 2\mathrm{Re}(\mathcal{S}^*\mathcal{P}) \propto 2|\mathcal{S}||\mathcal{P}|\cos(\xi_P + \delta_P - \xi_S - \delta_S) \tag{4}$$

$$\beta = 2\mathrm{Im}(\mathcal{S}^*\mathcal{P}) \propto 2|\mathcal{S}||\mathcal{P}|\sin(\xi_P + \delta_P - \xi_S - \delta_S) . \tag{5}$$

The real part given by the α parameter can be determined from the angular distribution of the baryon b when the baryon B has known non-zero polarization or by measuring the polarization of the daughter baryon. For example, the proton angular distribution in the $\Lambda(\Lambda \to p\pi^-)$ decay is given as

$$\frac{1}{\Gamma}\frac{\mathrm{d}\Gamma}{\mathrm{d}\Omega} = \frac{1}{4\pi}(1 + \alpha_\Lambda \mathbf{P}_\Lambda \cdot \hat{\mathbf{n}}) , \tag{6}$$

where $\mathbf{P}_\Lambda$ is the Λ polarization vector. Measurement of the imaginary part of the interference term given by the parameter β requires that the polarization of both baryon B and the daughter baryon is determined. For the decay $\Xi(\Xi^- \to \Lambda\pi^-)$, where the cascade is polarized, the β_Ξ parameter can be determined using the subsequent $\Lambda \to p\pi^-$ decay that acts as Λ polarimeter. The relation between the initial Ξ^- polarization $\mathbf{P}_\Xi$ and the daughter Λ polarization $\mathbf{P}_\Lambda$ is given by the Lee–Yang formula [11]. If the $\hat{\mathbf{z}}$ axis in the Ξ^- rest frame is defined along the $\hat{\mathbf{n}}$ direction than the relation between the polarization vectors is

$$\begin{bmatrix} P^x_\Lambda \\ P^y_\Lambda \\ P^z_\Lambda \end{bmatrix} = \frac{1}{1 + \alpha_\Xi P^z_\Xi} \begin{pmatrix} \gamma_\Xi P^x_\Xi - \beta_\Xi P^y_\Xi \\ \beta_\Xi P^x_\Xi + \gamma_\Xi P^y_\Xi \\ \alpha_\Xi + P^z_\Xi \end{pmatrix} , \tag{7}$$

where the parameter $\gamma = |\mathcal{S}|^2 - |\mathcal{P}|^2$. The equation implies that Λ in decay of unpolarized Ξ^- has the longitudinal polarization vector component $P^z_\Lambda = \alpha_\Xi$. Using a polar angle parameterization such that $\beta = \sqrt{1 - \alpha^2}\sin\phi$ and $\gamma = \sqrt{1 - \alpha^2}\cos\phi$, the ϕ parameter has interpretation of the rotation angle between the Ξ and Λ polarization vectors in the transversal x–y plane.

3 CP violation in baryon decays

For the C-transformed baryon decay process $\bar{B} \to \bar{b}\bar{\pi}$ the amplitude is:

$$\bar{\mathcal{A}}(\bar{B} \to \bar{b}\bar{\pi}) \propto \bar{\mathcal{S}}\sigma_0 - \bar{\mathcal{P}}\sigma \cdot \hat{\mathbf{n}} , \tag{8}$$

where the complex parameters $\bar{\mathcal{S}}$ and $\bar{\mathcal{P}}$ are obtained from $\mathcal{S}$ and $\mathcal{P}$ by reversing the sign for the weak CP-odd phases ξ_S and ξ_P. Since the product of the P-odd and P-even terms changes sign, the decay parameters are:

$$\bar{\alpha} \propto -2|\mathcal{S}||\mathcal{P}|\cos(-\xi_P + \delta_P + \xi_S - \delta_S) \quad (9)$$

$$\bar{\beta} \propto -2|\mathcal{S}||\mathcal{P}|\sin(-\xi_P + \delta_P + \xi_S - \delta_S)\,. \quad (10)$$

The weak phase difference $\xi_P - \xi_S$ can be determined using two independent experimental observables:

$$A_{CP} = \frac{\alpha + \bar{\alpha}}{\alpha - \bar{\alpha}} = (\xi_P - \xi_S)\tan(\delta_P - \delta_S),\; B_{CP} = \frac{\beta + \bar{\beta}}{\alpha - \bar{\alpha}} = (\xi_P - \xi_S)\,. \quad (11)$$

4 Spin entangled baryon–antibaryon systems

A novel method to study hyperon decays is to use the entangled baryon–antibaryon pair from J/ψ resonance decays. In general the spin state of a baryon–antibaryon system can be represented as

$$|B\bar{B}\rangle = \sum_{\mu,\nu=0}^{3} C_{\mu\nu}^{B\bar{B}} \sigma_{\mu}^{(B)} \otimes \sigma_{\nu}^{(\bar{B})}\,, \quad (12)$$

where the Pauli operators $\sigma_{\mu}^{(B)}$ and $\sigma_{\nu}^{(\bar{B})}$ act in the rest frames of the baryon and antibaryon, respectively. The coefficients of the spin correlation–polarization matrix $C_{\mu\nu}^{B\bar{B}}$ depend on the pair production process. Two reactions provide best sensitivity for the CP tests in hyperon decays: $e^+e^- \to J/\psi \to \Lambda\overline{\Lambda}$ [12, 13] and $e^+e^- \to J/\psi \to \Xi\overline{\Xi}$ [14]. At the BESIII experiment for every million of the produced J/ψ resonances, 320 and 56 of fully reconstructed $\Lambda\overline{\Lambda}$ and $\Xi\overline{\Xi}$ pairs are selected, respectively. With 10^{10} J/ψ mesons available at BESIII, precision measurements of the hyperon decay parameters are possible. The elements of the $C_{\mu\nu}^{B\bar{B}}$ matrix for the $e^+e^- \to B\overline{B}$ processes are known functions of the baryon B production angle θ and depend on two parameters that need to be determined from data [15, 16]. The baryons can be polarized in the direction perpendicular to the reaction plane given by the $\hat{\mathbf{y}}$ unit vector. The polarization vector component $P_y(\theta)$ and the spin correlation terms are $C_{ij}(\theta)$, $i, j = x, y, z$:

$$C_{\mu\nu}^{B\bar{B}} = \frac{1}{\sigma}\frac{\mathrm{d}\sigma}{\mathrm{d}\cos\theta}\begin{pmatrix} 1 & 0 & P_y & 0 \\ 0 & C_{xx} & 0 & C_{xz} \\ -P_y & 0 & C_{yy} & 0 \\ 0 & -C_{xz} & 0 & C_{zz} \end{pmatrix}. \quad (13)$$

The average baryon polarizations $|P_y|$ in the $e^+e^- \to J/\psi \to \Lambda\overline{\Lambda}$ and $e^+e^- \to J/\psi \to \Xi\overline{\Xi}$ reactions are 18% and 23%, respectively. The two production reaction parameters, the decay parameters α_Λ, α_Ξ and β_Ξ, as well as the CP-violating variables A_{CP}^{Λ}, A_{CP}^{Ξ} and B_{CP}^{Ξ}, can be determined using unbinned maximum likelihood fit to the measured angular distributions. The statistical uncertainty of the observable A_{CP}^{Λ} in the process $e^+e^- \to J/\psi \to \Lambda\overline{\Lambda}$ is inversely proportional to the Λ polarization [10]. The weak phase difference for the Ξ baryon decay is directly given by the measurement of the B_{CP}^{Ξ} observable in the process $e^+e^- \to J/\psi \to \Xi\overline{\Xi}$. The sensitivity of such measurement depends on the average squared of the Ξ polarization and the $\Xi\overline{\Xi}$ spin correlation terms [10].

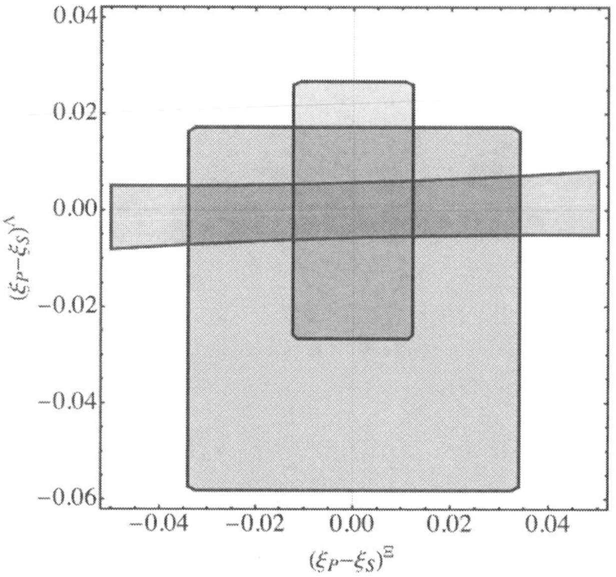

Figure 2. Summary of the results for the weak phases $(\xi_P - \xi_S)^\Xi$ and $(\xi_P - \xi_S)^\Lambda$. The present BESIII results are given by the blue rectangle [13, 14]. The red horizontal band is the HyperCP result [17]. The brown rectangle is the projection of the statistical uncertainties for the complete BESIII data set.

5 Weak phases

The goal of the CP tests is to determine the weak-phase differences for the two decays. The BESIII results based on 1.3×10^9 and 10^{10} J/ψ data for $(\xi_P-\xi_S)^\Xi$ [14] and $(\xi_P-\xi_S)^\Lambda$ [13], respectively, are presented in Fig. 2 by the blue rectangle. The red band represents the HyperCP experiment result on $A^\Lambda_{CP}+A^\Xi_{CP}$ [17]. Interpretation of this measurement requires the value of $\tan(\delta_P - \delta_S)_\Xi$ that is poorly known. Projection for the statistical uncertainties of the analysis using full BESIII data set for $e^+e^- \to J/\psi \to \Xi\bar{\Xi}$ is given by the brown rectangle. The results should be compared to the standard model predictions of $(\xi_P-\xi_S)^\Lambda = (-0.2\pm2.2)\times10^{-4}$ and $(\xi_P - \xi_S)^\Xi = (-2.1 \pm 1.7)\times 10^{-4}$. In case of hyperon decays the standard model contribution is dominated by QCD-penguin contributions shown in Fig. 1(a). The results on the weak phases in kaon and in the hyperon decays can be combined in order to search for deviations from the standard model. The present kaon data imply the limits $|\xi_P - \xi_S|^\Lambda_{\rm NP} \leq 5.3\times10^{-3}$ and $|\xi_P - \xi_S|^\Xi_{\rm NP} \leq 3.7\times10^{-3}$ [18]. Clearly, the hyperon CP-violation measurements with much improved precision will provide an independent constraint on the NP contributions in the strange quark sector. However, a lot also remains to be done on the theory side, as the present predictions suffer from considerable uncertainties. It is hoped that the lattice analyses [19] could help solve this problem in future.

This work was supported by the Polish National Science Centre through the Grant 2019/35-/O/ST2/02907.

References

[1] I.I. Bigi, A.I. Sanda, *CP violation*, Vol. 9 (Cambridge University Press, 2009), ISBN 978-0-511-57728-4, 978-0-521-84794-0, 978-1-107-42430-2
[2] M.S. Sozzi, Eur. Phys. J. C **36**, 37 (2004)

[3] R.L. Workman et al. (Particle Data Group), PTEP **2022**, 083C01 (2022)
[4] N. Cabibbo, Phys. Rev. Lett. **10**, 531 (1963)
[5] M. Kobayashi, T. Maskawa, Prog. Theor. Phys. **49**, 652 (1973)
[6] L. Wolfenstein, Phys. Rev. Lett. **51**, 1945 (1983)
[7] H. Gisbert, A. Pich, Rept. Prog. Phys. **81**, 076201 (2018)
[8] Z. Bai et al. (RBC, UKQCD), Phys. Rev. Lett. **115**, 212001 (2015)
[9] R. Abbott et al. (RBC, UKQCD), Phys. Rev. D **102**, 054509 (2020)
[10] N. Salone, P. Adlarson, V. Batozskaya, A. Kupsc, S. Leupold, J. Tandean, Phys. Rev. D **105**, 116022 (2022)
[11] T.D. Lee, C.N. Yang, Phys. Rev. **108**, 1645 (1957)
[12] M. Ablikim et al. (BESIII), Nature Phys. **15**, 631 (2019)
[13] M. Ablikim et al. (BESIII), Phys. Rev. Lett. **129**, 131801 (2022)
[14] M. Ablikim et al. (BESIII), Nature **606**, 64 (2022)
[15] G. Fäldt, A. Kupsc, Phys. Lett. **B772**, 16 (2017)
[16] E. Perotti, G. Fäldt, A. Kupsc, S. Leupold, J.J. Song, Phys. Rev. D **99**, 056008 (2019)
[17] T. Holmstrom et al. (HyperCP), Phys. Rev. Lett. **93**, 262001 (2004)
[18] X.G. He, H. Murayama, S. Pakvasa, G. Valencia, Phys. Rev. D **61**, 071701 (2000)
[19] S.R. Beane, P.F. Bedaque, A. Parreno, M.J. Savage, Nucl. Phys. A **747**, 55 (2005)

Search for CP-Violation in ortho-Positronium Decay

Tom-Erik Haugen[1,*], *Elizabeth A.* George[2], *Oscar* Naviliat-Cuncic[1,3], and *Paul A.* Voytas[2]

[1]Facility for Rare Isotope Beams, and Department of Physics and Astronomy, Michigan State University, East Lansing, Mi, USA
[2]Wittenberg University, Springfield, Oh, USA
[3]Laboratoire de Physique Corpusculaire de Caen, and Université de Caen Normandie, Caen, France

Abstract. We report here on the design and development of a new apparatus aiming at a 10-fold improved search for CP-violation in ortho-Positronium decay to be performed at the Facility for Rare Isotope Beams.

1 Positronium

Positronium (Ps) is a Hydrogen-like system comprised of a bound positron and electron. It is a purely leptonic bound state making it insensitive to hadronic physics or strong interactions. It is an eigenstate of the charge conjugation operator (C), and the Parity operator (P), where the C and P eigenvalues depend on the spin configuration and principal quantum numbers. Positronium constitutes a clean system to search for CP-violation, and any observed CP-violation would be due to Beyond Standard Model physics. Positronium has two spin configurations, para-positronium (p-Ps) the spin-zero singlet state, and ortho-positronum (o-Ps) the spin-one triplet state. By charge selection rules p-Ps can only decay to an even number of photons (primarily 2), and o-Ps can only decay to an odd number (primarily 3). This leads to a much larger lifetime for o-Ps (142 ns) compared to p-Ps (125 ps).

1.1 Positronium Decay

We order the γ's in the 3-γ decay of o-Ps by their energies as indicated in figure 1a). The energy distribution and angular distribution of photons from unpolarized ortho-positronium decay is given in figure 1c) and 1d) respectively. Due to rotational invariance the angular distribution of the decay photons must be proportional to products of the γ momenta and the initial o-Ps spin. We will search for a correlation of the form,

$$\left(\hat{\mathbf{s}} \cdot \hat{\mathbf{k}}_i\right) \hat{\mathbf{s}} \cdot \left(\hat{\mathbf{k}}_1 \times \hat{\mathbf{k}}_2\right) \tag{1}$$

where $\mathbf{s}$ is the spin of the o-Ps, $\mathbf{k}_i$ is the momentum vector for one of the photons. Such a correlation can only be induced by CP-violating physics [1]. Two previous experiments have placed limits on the size of such a correlation of 0.01 in [2], and 0.001 in [3]. We are designing an apparatus to reach a precision of 10^{-4}. We will measure events where we detect both the highest and second highest energy photons. With a highly symmetric detector array we can flip the sign of the correlation and record differences in counts that will be directly sensitive to CP-violating physics.

*e-mail: haugen@frib.msu.edu

 https://doi.org/10.1051/epjconf/202328201003

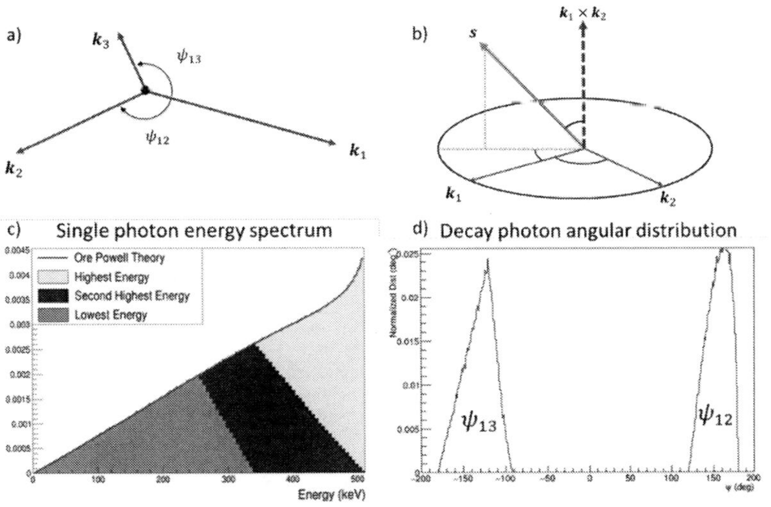

Figure 1. a) Ordering of decay photons by their energies. b) Relevant vectors in o-Ps decay. c) Energy distribution for photons from unpolarized o-Ps. d) Angular distribution of photons within the decay plane for unpolarized o-Ps.

1.2 Positronium Alignment

This correlation requires the use of tensor-polarized positronium and is proportional to the alignment of the o-Ps given as,

$$P_2 = \frac{N_+ - 2N_0 + N_-}{N_+ + N_0 + N_-} \tag{2}$$

where N_i are the population of o-Ps states. We can achieve a nonzero tensor polarization with the same method used in [2]. We utilize a magnetic field to mix the singlet state with the $m = 0$ triplet state. This mixed "pseudo-triplet" state has a shorter lifetime than that of the $N_\pm$ and so P_2 changes over time. The setup includes a cryocooled superconducting magnet which produces a maximum field of 2 T. For a modest field value of 0.4 T the pseudo-triplet lifetime is reduced to 25 ns. Integrating the events within two time windows will give two populations with opposite values of alignment.

2 Positronium Formation

We have constructed a test stand to demonstrate and optimize Ps formation. We use a ^{22}Na source surrounded by finely grained powder. The β^+ slows down while traveling through the powder, and can pick up an electron and form positronium. With our test stand we detect the 1275 keV γ-ray from the ^{22}Na decay, and one of the photons from the Ps annihilation. We use a PIXIE-16 module and NSCLDAQ to read out two $LaBr_3$ detectors in coincidence. A sample time spectrum is shown in figure 2.

Future work includes baking the powder and pumping under vacuum to increase the lifetime of o-Ps by removing moisture. We plan to achieve a larger formation fraction by testing different powders (MgO, SiO_2, etc.) as well as different grain sizes for the powders.

https://doi.org/10.1051/epjconf/202328201003

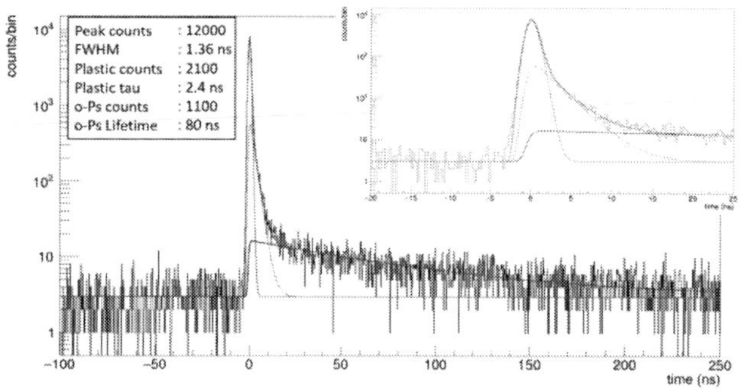

Figure 2. Demonstration of Positronium formation. Early studies clearly show a sharp prompt peak, a component of Ps forming in the plastic surrounding the source, and a long lifetime component of Ps in the MgO powder. The data was taken with 20 nm grain size MgO powder at 0.3 g/cm^3 density and a comercial ^{22}Na source with an activity of 0.16 MBq.

3 Photon Detectors

The annihilation radiation will be detected using crystal scintillators. Lutetium Yttrium Orthosilicate (LYSO) is a crystal scintillator with high scintillation light yield, and moderately fast decay time (45 ns). We have extensively studied LYSO crystals coupled to Silicon Photomultipliers (SiPM). With the prototype detectors we have achieved an energy resolution of 13% at 511 keV using a prototype crystal in the tapered geometry, shown in figure 3. With a pair of crystals we have achieved 1.87 ns timing resolution for coincidences. This will be vital for reducing the rate of random coincidences. We will impose a triple coincidence trigger online, requiring hits in two LYSO crystals and a start coming from a plastic scintillator that the β^+ crosses on its way to the Ps formation region. Due to the compactness of our detectors and the SiPM's we will be able to fit a large number of detectors inside the warm bore of the magnet with a diameter of 22 cm.

Figure 3. A prototype LYSO crystal readout by a Silicon photomultiplier. The tapered design was selected to cover a larger solid angle, and increase photopeak efficiency compared to a parallelepiped. The crystal is 3 cm along the long axis. This picture does not feature any crystal wrapping.

4 Full Apparatus Design

The full apparatus will consist of 3 rings of γ detectors, with 16 detectors in each ring as depicted in figure 4. Due to the angular distribution of k_1 and k_2 shown in figure 1d), for an event where k_1 hits the detector marked with green in figure 4a), k_2 from those decays are detected mostly by our detectors in the opposite ring marked in red and blue. For the process just described the upper red and blue detectors have opposite signs of the angular correlations from their symmetrically located partners (the lower red and blue marked detectors). CP violation would manifest itself as a difference in the number of counts for the upper vs. lower combinations. Similar pairings around the axis of symmetry, as well as other combinations between the outer rings and the center ring, greatly increase the number of detector combinations and solid angle coverage giving us an improved statistical reach compared to previous experiments.

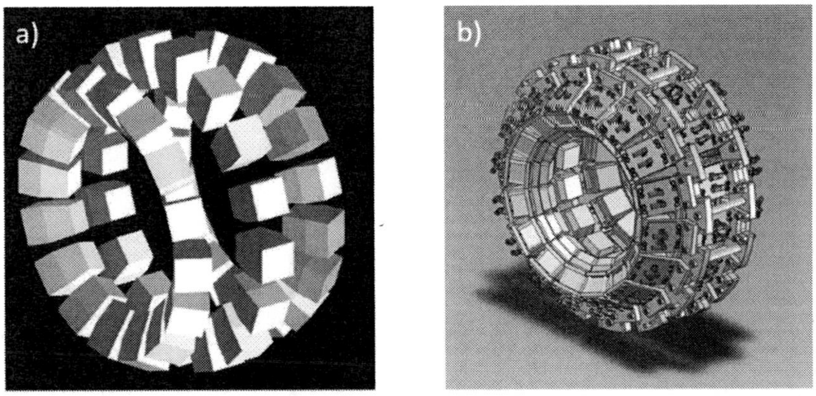

Figure 4. a) Full detector array geometry, placement has been optimized for sensitivity to our observable. b) Support structure to hold all γ- detectors and readout electronics within the warm bore of our magnet.

4.1 Simulations

The geometry has been optimized in concurrent simulations using both Geant4 and EGSnrc. This required writing a custom 3-γ event generator. Effects studied include, 2-γ dilution, random coincidences mimicking signal events, effects of energy resolution and energy cuts, and spreading of the source beyond a point. Our simulations showed no benefit from inclusion of shielding between rings of detectors. With our finalized detector designs, and optimized geometry we have finished design of the full detector array including the support structure as shown in figure 4b).

5 Summary

We have designed an apparatus to search for CP-violation in o-Ps decay. Our compact and highly symmetric design is able to fit within the warm bore of our magnet. We have studied and prototyped our photon detectors, demonstrated formation of Ps, and simulated possible sensitivity limiting effects. Future work includes optimization of Ps formation, construction of the apparatus, and continued study of systematic effects. This design will allow for us to reach a 10 x higher sensitivity than previous searches.

This material is based upon work supported by the National Science Foundation under Grants No. PHY-1920065, PHY-2013557, PHY-1565546, PHY-2111185, and by the U.S. Department of Energy, Office of Science, Office of Nuclear Physics, and used resources of the Facility for Rare Isotope Beams (FRIB), which is a DOE Office of Science User Facility, operated by Michigan State University, under Award Number DE-SC0000661.

References

[1] W. Bernreuther, U. Löw, J.P. Ma, et al. Z. Phys. C - Particles and Fields **41**, pp 143–158 (1988).

[2] M. Skalsey and J. Van House, Phyr. Rev. Lett. **67**, pp 1993-1996 (1991)

[3] T. Yamazaki, T. Namba, S. Asai, T. Kobayashi, Phys. Rev. Lett. **104**, Article 083401 (2010)

EPJ Web of Conferences **282**, 01004 (2023) https://doi.org/10.1051/epjconf/202328201004
SSP 2022

The Standard-Model Extension

Ralf Lehnert[1,*]

[1]Indiana University Center for Spacetime Symmetries, Bloomington, IN 47405, USA

Abstract. Lorentz and CPT symmetry represent cornerstones of our present understanding of nature, but may be violated in various theoretical approaches to underlying physics. Testing these symmetries therefore establishes a promising avenue to search for physics beyond the Standard Model. The canonical theoretical tool to identify possible experimental signatures of such violations is an effective-field-theory framework known as the Standard-Model Extension. This talk provides an overview of this topic with focus on efforts involving low-energy atomic and subatomic systems.

1 Introduction

The concept of symmetries—and their breakdown—plays a key role in understanding, modeling, testing, and advancing physics. Due to their relative simplicity, atomic and subatomic systems are exceptionally well positioned for investigations of various discrete and continuous symmetries, and the ultra-high sensitivity attainable in these systems offers real discovery potential for physics beyond the Standard Model (SM).

Lorentz and CPT invariance are no exception to this assessment: past atomic and subatomic tests of these spacetime symmetries are among the most precise ones available with some of them reaching sensitivities that may be interpreted as probing the Planck scale [1]. Conceptual motivations for such studies arise from theoretical ideas that seek to answer questions beyond established physics because they often accommodate small deviations from Lorentz and CPT symmetry [2–6].

For the identification and interpretation of currently feasible Lorentz and CPT tests a theoretical framework has been developed. This framework, called the Standard-Model Extension (SME), classifies all possible perturbative departures from these spacetime symmetries that are compatible with coordinate independence and standard effective field theory [7]. All measurements of Lorentz and CPT symmetry tabulated in the aforementioned Ref. [1] are given as constraints on SME coefficients.

The outline of this work is as follows. A brief overview of SME basics is contained in Sec. 2. Section 3 discusses as an example a recent Lorentz test in a subatomic system, namely Tritium beta decay. Measurements of Lorentz and CPT symmetry in an atomic system, (anti)hydrogen, are the topic of Sec. 4. Throughout this work, we employ natural units $\hbar = c = k_B = 1$.

*e-mail: ralehner@indiana.edu

2 SME Basics

In view of the various theoretical ideas for Lorentz and CPT violation in underlying physics cited in the introduction and the ambitious goal to cover a broad range of experimental searches for these effects, the SME test framework needs to describe departures from Lorentz and CPT symmetry with great generality at presently attainable energies. To this end, the SME is constructed by hand rather than derived as the low-energy limit of one of the underlying models incorporating Lorentz and CPT breakdown. At the same time, the SME maintains as many other fundamental physics principles as possible. What follows is an overview of the key ideas behind the construction of the SME framework.

One of these ideas is an effective field theory (EFT) approach. This approach is widely—and with great success—employed in a number of different areas in physics such as elementary-particle, nuclear, and condensed-matter physics. Moreover, it is theoretically well understood and provides a general and flexible description of dynamical systems with large numbers of degrees of freedom. It is thus a natural assumption that the low-energy leading-order physics of a potential breakdown of Lorentz and CPT symmetry originating in underlying physics can be described within EFT. This assumption leads to a lagrangian formulation of the SME as an EFT.

A second key idea concerns the characterization of departures from Lorentz and CPT symmetry. A common feature of all of the aforementioned sample mechanisms for Lorentz and CPT violation is the emergence of preferred directions in the vacuum; these are the physical objects causing the symmetry breakdown. Based on this insight, Lorentz and CPT breaking is parametrized by non-dynamical external vector or tensor fields $b^\mu, c^{\mu\nu}, \ldots$ in the SME. In the flat-spacetime limit, these are usually taken to be constant ensuring energy–momentum conservation. Note the analogy to the spontaneous breakdown of a conventional gauge symmetry, which also generates a constant background field carrying the indices of some representation of the corresponding symmetry group.

A third key idea is coordinate independence: although Lorentz and CPT invariance are violated, it should still be possible to select any suitable reference frame for the mathematical description of physical laws. In other words, coordinates are a pure product of human thought and should not acquire physical significance. Coordinate independence is guaranteed if physics is formulated in terms of geometrical quantities, such as scalars, vectors, tensors, and spinors. This implies that the SME background vectors and tensors $b^\mu, c^{\mu\nu}, \ldots$ must enter the SME Lagrangian with all their indices contracted. Note again the analogy to the spontaneous breakdown of a conventional gauge symmetry: Each term in the Lagrangian is a gauge singlet with all group indices contracted. This features is then trivially maintained, when one object in such a term acquires a vacuum expectation value.

These ideas inspire the following general structure of the SME Lagrangian $\mathcal{L}_{\text{SME}}$:

$$\mathcal{L}_{\text{SME}} = \mathcal{L}_{\text{SM}} + \mathcal{L}_{\text{EH}} + \delta\mathcal{L}_{\text{SME}}\,, \tag{1}$$

where $\mathcal{L}_{\text{SM}}$ and $\mathcal{L}_{\text{EH}}$ are the ordinary SM and Einstein–Hilbert pieces, and $\delta\mathcal{L}_{\text{SME}}$ contains small Lorentz- and CPT-violating corrections constructed according to the reasoning above:

$$\delta\mathcal{L}_{\text{SME}} = -\overline{\psi} b^\mu \gamma_5 \gamma_\mu \psi + i\overline{\psi} c^{\mu\nu} \gamma_\mu \partial_\nu \psi + \ldots\,, \tag{2}$$

where b^μ and $c^{\mu\nu}$ are the aforementioned examples for preferred directions assumed to be generated by underlying physics. Within the present context, they represent phenomenological coefficients that govern the type and extent of Lorentz violation. We remark that the b^μ coefficient also controls certain types of CPT breaking, while $c^{\mu\nu}$ is CPT even.

Other desirable physics features, such as the ordinary $U(1) \times SU(2) \times SU(3)$ gauge symmetry, are usually imposed on the SME. The overview above has primarily focused on the flat-spacetime SME. However, these ideas can be generalized to curved-spacetime situations involving gravity. We finally remark that the SME has provided the basis for numerous theoretical and mathematical investigations of Lorentz and CPT violation [8–13], and that the framework has also found applications in adjacent research fields [14].

As an EFT, the SME lends itself to a natural classification of Lorentz- and CPT-violating contributions according to their mass dimension d. The restriction to $d = 3, 4$ is usually called the minimal SME (mSME), and it is expected that this restriction contains the effects dominant at low energies. However, for broadest possible generality, the full SME also contains nonminimal contributions $d > 4$. For example, specific underlying models, such as noncommutative field theory, are indeed known to generate Lorentz-violating corrections starting at $d = 5$. This classification can be refined in various ways including by considering the C, P, and T properties of a given SME term.

3 Tritium beta decay

The neutrino remains one of the least understood particle species, and there is correspondingly large experimental and theoretical interest in neutrino investigations including Lorentz- and CPT-symmetry studies. Measurements involving flavor oscillations are interferometric in nature and thus offer unparalleled sensitivity to the neutrino's SME coefficients [1]. The long propagation distances of astrophysical neutrinos can also lead to tight constraints on Lorentz and CPT violation in the neutrino sector via time-of-flight measurements [1].

There are, however, some SME effects that leave both the oscillation pattern and the neutrino's velocity unaffected. The corresponding coefficients are termed 'countershaded.' The difficulty of accessing them experimentally leaves open the possibility of relatively large countershaded SME contributions that may have thus far evaded detection. Such effects therefore need to be searched for using other observables and experimental techniques.

One type of these countershaded contributions is generated by the SME's a^μ coefficients. They are associated to fermions and may have independent values for each species. In situations in which gravity can be disregarded, these coefficients generate a constant shifts by a^μ in the particle's 4-momentum and are known to be undetectable in systems involving only a single species of fermions. For this reason, previous experimental studies have employed systems with multiple species of fermions [15–21] or gravitational backgrounds [22, 23], and they have constrained a^μ coefficients for various quark flavors as well as the proton and the electron [1].

The SME's a^μ coefficients are also known to alter the phase space available to the outgoing fermions in particle reaction processes. In this context, the β-decay spectrum in particular has been the topic of prior theoretical studies [24, 25]: it involves an outgoing neutrino and thus provides experimental access to this fermion's a^μ coefficient. Moreover, the endpoint energy of this decay process has already been measured in various high-precision experimental investigations, and continuing efforts in this context are slated for future improvements in sensitivity [26–33]. What follows is a brief review of the most recent update of this particular type of measurement at KATRIN [34, 35].

The KATRIN experiment measures the β spectrum of Tritium decay

$$ {}^3_1\mathrm{T} \to {}^3_2\mathrm{He}^+ + e^- + \bar{\nu}_e \,. \tag{3} $$

The dynamics of this low-energy decay is primarily governed by the involved two nuclei, the electron, and the neutrino. Correspondingly, we treat all these particles effectively as spin-$\frac{1}{2}$ fermions, as usual [36]. Countershaded a^μ Lorentz and CPT breaking in this system is then given by the lagrangian contributions for each of these four particles

$$\delta\mathcal{L}^{a}_{\text{SME}} = -\bar{\psi}_w\, a^{\mu}_w \gamma_\mu \psi_w\,. \tag{4}$$

Here, the species subscript $w \in \{T, H, e, n\}$ denotes the tritium, the helium, the electron, and the neutrino, respectively. There are also numerous other SME coefficients that can affect the decay (3), but many of these have already been bounded in other physical systems at sensitivities that render them effectively zero in the present context.

It turns out that in the absence of interactions, all four a^μ-type coefficients can be removed from the Lagrangian via a field redefinition, so that they would be undetectable [7]. However, this reasoning no longer holds when there are interactions. One can show that in the present situation, for example, only

$$a^\mu \equiv a^\mu_T - a^\mu_H - a^\mu_e + a^\mu_n \tag{5}$$

is observable [34]. This fact represents one of the reasons why this type of SME effect is countershaded.

With this input the leading-order a^μ correction to the conventional tree-level shape of the β spectrum can be determined [34]. A particularly useful observable is its endpoint energy. In general, it depends on the direction of the emitted electron due to the presence of a^μ. The actual KATRIN setup involves the collection of electrons emitted in all directions inside an acceptance cone of opening angle κ and pointing in direction $\hat{c}$, precluding the notion of a unique, single endpoint. However, averaging over the momenta inside the cone, one may introduce a 'mean' endpoint

$$\bar{E}^a_m(\hat{c}) \equiv E_m + a^0 + M_T^{-1}\sqrt{E_m^2 - m_e^2}\,\hat{c}\cdot\vec{a}\,\cos^2\tfrac{1}{2}\kappa \tag{6}$$

valid at first order in a^μ. Here, we have linearized in E_e by dropping terms of order $E_m(E_m - E_e)/(E_m^2 - m_e^2)^{1/2}$. The conventional endpoint is denoted by E_m, and the Tritium mass is M_T. Intuitively, $\bar{E}^a_m(\hat{c})$ represents the endpoint corresponding to the average cone momentum and therefore depends on the cone direction $\hat{c}$.

As a terrestrial experiment, KATRIN's orientation and velocity is time dependent due to the motion of the Earth. On the other hand, a^μ is presumed to be spacetime constant. Result (6) therefore predicts in general a time-dependent mean endpoint energy $\bar{E}^a_m$. Focusing solely on the orientation change due to the Earth's rotation about its axis, the direction of the cone vector $\hat{c} = \hat{c}(t)$ will exhibit a periodic time dependence determined by the length of a sidereal day. For this situation, Result (6) forecasts a constant shift by a^0 in the endpoint energy relative to its conventional value, as well as sinusoidal variations of the endpoint with a sidereal-day period and an amplitude governed by $\vec{a}$.

Available time-stamped KATRIN data together with information on KATRIN's location and orientation on Earth can be used to perform an experimental search for such effects. Such an analysis has indeed been recently performed by KATRIN leading to a constraint on a^μ at the level of 10^{-6} GeV [35]. This result represents the world's first, and thus best, measurement of this type.

4 Tests involving antihydrogen

The availability of slow antiprotons at CERN's AD facility has permitted the production of cold antihydrogen in recent years. This achievement, in turn, has opened an avenue for experimental hydrogen–antihydrogen comparisons, and thus atomic tests of CPT symmetry. This sections lists some ongoing efforts in this field within the SME context.

The lagrangian SME terms $\delta\mathcal{L}_{\rm SME}$ lead for example to contributions $\delta H_{\rm SME}$ to one-particle Hamiltonians, which in turn generate corrections $\delta E_{\rm SME}$ to the energy spectrum of bound states. Spectroscopic investigations of atomic spectra therefore represent a well-suited experimental tool for ultra-high precision Lorentz and CPT tests. Paralleling the conceptual underpinnings of β-decay studies discussed in the previous sections, one type of atomic-spectroscopy measurements of SME coeffcients monitors transition frequencies over time and analyzes them for sidereal variations. But the availability of matter–antimatter conjugates for hydrogen also allows the instantaneous comparison of a given transition in H with that in $\overline{\rm H}$. Such a direct comparison is typically sensitive to those SME coefficients that violate both Lorentz and CPT symmetry.

Among the transitions in (anti)hydrogen that are affected at leading order is the ground-state hyperfine splitting. In H, experimentally attainable relative sensitivities to these splittings are on the order of 10^{-12}. A correspondingly high sensitivity in $\overline{\rm H}$ would therefore allow an excellent test of CPT invariance. An experiment along these lines is currently being performed by the ASACUSA collaboration at CERN's AD/ELENA facility [37]. The goal is an in-flight measurement of the π transition $(1, 1) \to (0, 0)$. Here, the expressions (F, M_F) denote a state with total angular momentum F and projection M_F onto the trapping magnetic field. Within the mSME, this transition can acquire a shift $\delta\nu$ in H and a different shift $\delta\bar{\nu}$ in $\overline{\rm H}$ with [38]

$$\delta\nu - \delta\bar{\nu} = -\frac{1}{\pi}\left(b_3^e - m_e g_3^{e(A)} + m_e g_{120}^{e(M)} + b_3^p - m_p g_3^{p(A)} + m_p g_{120}^{p(M)}\right) + \ldots\,, \tag{7}$$

where m_e and m_p are the electron and proton mass, respectively, and the various b and g quantities are certain components of CPT-violating SME coefficients. The ellipses indicate nonminimal SME contributions.

Measurements of the 1S–2S transition represent another avenue for a Lorentz and CPT test in H and $\overline{\rm H}$. In H, this transition can be measured with a relative precision at the 10^{-15} level. As for the aforementioned HF transition, this number bodes well for a H–$\overline{\rm H}$ comparison with ultra-high precision. Such a type of measurement is the goal of the ALPHA collaboration, which is also located at CERN's AD/ELENA facility [39]. For this situation, an SME analysis reveals a transition-frequency correction $\delta\nu_{1S2S}$ in H and again a different correction $\delta\bar{\nu}_{1S2S}$ in $\overline{\rm H}$ with [38]

$$\delta\nu_{1S2S} - \delta\bar{\nu}_{1S2S} = -\frac{8}{\pi}\sum_w \left[12(\alpha m_r)^2\, \mathring{a}^{\rm NR}_{w,2} + 67(\alpha m_r)^4\, \mathring{a}^{\rm NR}_{w,4}\right]. \tag{8}$$

In this expression, α denotes the fine-structure constant and m_r the reduced mass. The species label w runs over the values e and p, and $\mathring{a}^{\rm NR}_{w,2}$ and $\mathring{a}^{\rm NR}_{w,4}$ are SME coefficients. We remark that there are no leading-order effects predicted within the mSME; the $\mathring{a}^{\rm NR}_{w,2}$ and $\mathring{a}^{\rm NR}_{w,4}$ coefficients are part of the nonminimal SME.

Another experimental question that can be addressed with $\overline{\rm H}$ concerns the interaction of antimatter with gravity. Various efforts are currently underway to perform such a measurement including AEGIS [40], GBAR [41], ALPHA [42], and AGE [43]. These efforts are focused in particular on studying the free fall of $\overline{\rm H}$ in Earth's gravitational field. Since the gravitationally coupled SME contains Lorentz- and CPT-violating interactions that can affect free-fall trajectories, such experimental efforts are well positioned to place constraints on the corresponding SME coefficients. One particular isotropic SME effect consists of a correction

$$\delta F_z = -2g\alpha\left(\overline{a}_t^{\rm T} + \overline{a}_t^{\rm S}\,\frac{m^{\rm T}}{m^{\rm S}}\right) \tag{9}$$

to the gravitational force exerted by the source mass m^{S}, taken to be Earth, on the test mass m^{T} [22]. Here, α is a parameter that captures dynamical effects of the underlying physics generating Lorentz and CPT violation, the two $\overline{a}$ quantities are SME coefficients, and g is the usual gravitational acceleration on Earth.

References

[1] V.A. Kostelecký and N. Russell, *Data Tables for Lorentz and CPT Violation*, 2022 edition, arXiv:0801.0287v15.

[2] V.A. Kostelecký and S. Samuel, Phys. Rev. D **39**, 683 (1989); V.A. Kostelecký and R. Potting, Nucl. Phys. B **359**, 545 (1991); Phys. Rev. D **51**, 3923 (1995); K. Hashimoto and M. Murata, PTEP **2013**, 043B01 (2013).

[3] I. Mocioiu, M. Pospelov, and R. Roiban, Phys. Lett. B **489**, 390 (2000); S.M. Carroll *et al.*, Phys. Rev. Lett. **87**, 141601 (2001); C.E. Carlson, C.D. Carone, and R.F. Lebed, Phys. Lett. B **518**, 201 (2001); A. Anisimov *et al.*, Phys. Rev. D **65**, 085032 (2002).

[4] J. Alfaro, H.A. Morales-Técotl, and L.F. Urrutia, Phys. Rev. Lett. **84**, 2318 (2000); F.R. Klinkhamer and C. Rupp, Phys. Rev. D **70**, 045020 (2004); N.E. Mavromatos, Lect. Notes Phys. **669**, 245 (2005).

[5] V.A. Kostelecký, R. Lehnert, and M.J. Perry, Phys. Rev. D **68**, 123511 (2003); R. Jackiw and S.Y. Pi, Phys. Rev. D **68**, 104012 (2003); O. Bertolami *et al.*, Phys. Rev. D **69**, 083513 (2004); N. Arkani-Hamed *et al.*, JHEP **07**, 029 (2005).

[6] F.R. Klinkhamer, Nucl. Phys. B **578**, 277 (2000); C.P. Burgess *et al.*, JHEP **03**, 043 (2002); J.D. Bjorken, Phys. Rev. D **67**, 043508 (2003); D. Grumiller, P. Parekh, and M. Riegler, Phys. Rev. Lett. **123**, 121602 (2019).

[7] D. Colladay and V.A. Kostelecký, Phys. Rev. D **55**, 6760 (1997); Phys. Rev. D **58**, 116002 (1998); V.A. Kostelecký, Phys. Rev. D **69**, 105009 (2004); V.A. Kostelecký and M. Mewes Phys. Rev. D **80**, 015020 (2009); Phys. Rev. D **85**, 096005 (2012); Phys. Rev. D **88**, 096006 (2013); V.A. Kostelecký and Z. Li, Phys. Rev. D **99**, 056016 (2019); Phys. Rev. D **103**, 024059 (2021).

[8] R. Bluhm and V.A. Kostelecký, Phys. Rev. D **71**, 065008 (2005); R. Bluhm *et al.*, Phys. Rev. D **77**, 125007 (2008) [erratum: Phys. Rev. D **79**, 029902 (2009)]; M. Cambiaso, R. Lehnert, and R. Potting, Phys. Rev. D **85**, 085023 (2012); C.A. Escobar and R. Potting, Int. J. Mod. Phys. A **35**, 2050174 (2020).

[9] V.A. Kostelecký and R. Lehnert, Phys. Rev. D **63**, 065008 (2001); V.A. Kostelecký, C.D. Lane, and A.G.M. Pickering, Phys. Rev. D **65**, 056006 (2002); O.W. Greenberg, Phys. Rev. Lett. **89**, 231602 (2002); D. Colladay and P. McDonald, Phys. Rev. D **77**, 085006 (2008); Phys. Rev. D **79**, 125019 (2009); B. Altschul, Q.G. Bailey, and V.A. Kostelecký, Phys. Rev. D **81**, 065028 (2010); F.R. Klinkhamer and M. Schreck, Nucl. Phys. B **848**, 90 (2011); M. Cambiaso, R. Lehnert, and R. Potting, Phys. Rev. D **90**, 065003 (2014); M.M. Ferreira *et al.*, Phys. Rev. D **100**, 055036 (2019); A.F. Ferrari, J.R. Nascimento, and A.Y. Petrov, Eur. Phys. J. C **80**, 459 (2020).

[10] D. Colladay and P. McDonald, J. Math. Phys. **43**, 3554 (2002); R. Lehnert, J. Math. Phys. **45**, 3399 (2004); Phys. Rev. D **74**, 125001 (2006); B. Altschul, J. Phys. A **39**, 13757 (2006); A.J. Hariton and R. Lehnert, Phys. Lett. A **367**, 11 (2007); R. Lehnert, Rev. Mex. Fis. **56**, 469 (2010); J.A.A.S. Reis and M. Schreck, Phys. Rev. D **95**, 075016 (2017).

[11] J.P. Noordmans, J. de Vries, and R.G.E. Timmermans, Phys. Rev. C **94**, 025502 (2016); J.P. Noordmans, Phys. Rev. D **95**, 075030 (2017); R. Kamand, B. Altschul,

and M.R. Schindler, Phys. Rev. D **95**, 056005 (2017); Phys. Rev. D **97**, 095027 (2018); B. Altschul and M.R. Schindler, Phys. Rev. D **100**, 075031 (2019).

[12] Q.G. Bailey and V.A. Kostelecký, Phys. Rev. D **74**, 045001 (2006); V.A. Kostelecký and J. D. Tasson, Phys. Rev. D **83**, 016013 (2011); Y. Bonder, Phys. Rev. D **91**, 125002 (2015); Y. Bonder and C.A. Escobar, Phys. Rev. D **93**, 025020 (2016); K. O'Neal-Ault, Q.G. Bailey, and N.A. Nilsson, Phys. Rev. D **103**, 044010 (2021); C.M. Reyes and M. Schreck, arXiv:2105.05954.

[13] V.A. Kostelecký, Phys. Lett. B **701**, 137 (2011); V.A. Kostelecký, N. Russell, and R. Tso, Phys. Lett. B **716**, 470 (2012); J. Foster and R. Lehnert, Phys. Lett. B **746**, 164 (2015); D. Colladay and P. McDonald, Phys. Rev. D **92**, 085031 (2015); M. Schreck, Phys. Rev. D **93**, 105017 (2016).

[14] V.A. Kostelecký, N. Russell, and J. Tasson, Phys. Rev. Lett. **100**, 111102 (2008); R. Lehnert, W.M. Snow, and H. Yan, Phys. Lett. B **730**, 353 (2014) [erratum: Phys. Lett. B **744**, 415 (2015)]; J. Foster, V.A. Kostelecký, and R. Xu, Phys. Rev. D **95**, 084033 (2017); R. Lehnert *et al.*, Phys. Lett. B **772**, 865 (2017); A. Martín-Ruiz, M. Cambiaso, and L.F. Urrutia, Phys. Rev. B **99**, 155142 (2019); V.A. Kostelecký *et al.*, Phys. Rev. Res. **4**, 023106 (2022).

[15] V.A. Kostelecký, Phys. Rev. D **61**, 016002 (2000); Phys. Rev. D **64**, 076001 (2001); Á. Roberts, Phys. Rev. D **96**, 116015 (2017).

[16] Y.B. Hsiung [KTeV], Nucl. Phys. B Proc. Suppl. **86**, 312 (2000); H. Nguyen [KTeV], in V.A. Kostelecký, ed., *CPT and Lorentz Symmetry II*, World Scientific, Singapore, 2002 [hep-ex/0112046].

[17] J.M. Link *et al.* [FOCUS], Phys. Lett. B **556**, 7 (2003).

[18] B. Aubert *et al.* [BaBar], Contribution to ICHEP 2006 [arXiv:hep-ex/0607103]; B. Aubert *et al.* [BaBar], Phys. Rev. Lett. **100**, 131802 (2008); K.R. Schubert, in V.A. Kostelecký, ed., *CPT and Lorentz Symmetry VII*, World Scientific, Singapore, 2017 [arXiv:1607.05882].

[19] F. Bossi *et al.* [KLOE], Riv. Nuovo Cim. **31**, 531 (2008); M. Testa [KLOE], in J.-M. Frère, L. Iconomidou-Fayard, S. Loucatos, J. Trân Thanh Vân, eds., *43rd Rencontres de Moriond on Electroweak Interactions and Unified Theories*, Thê Gioi Publishers, 2008 [arXiv:0805.1969]; A. Di Domenico [KLOE], in V.A. Kostelecký, ed., *CPT and Lorentz Symmetry IV*, World Scientific, Singapore, 2008; J. Phys. Conf. Ser. **171**, 012008 (2009); Found. Phys. **40**, 852-866 (2010).

[20] V.A. Kostelecký and R. Van Kooten, Phys. Rev. D **82**, 101702 (2010); V.M. Abazov *et al.* [D0], Phys. Rev. Lett. **115**, 161601 (2015).

[21] J. van Tilburg and M. van Veghel, Phys. Lett. B **742**, 236 (2015); R. Aaij *et al.* [LHCb], Phys. Rev. Lett. **116**, 241601 (2016).

[22] V.A. Kostelecky and J. Tasson, Phys. Rev. Lett. **102**, 010402 (2009); Phys. Rev. D **83**, 016013 (2011).

[23] M.A. Hohensee *et al.*, Phys. Rev. Lett. **106**, 151102 (2011).

[24] J.S. Díaz, V.A. Kostelecký, and R. Lehnert, Phys. Rev. D **88**, 071902 (2013).

[25] J.S. Díaz, Phys. Rev. D **89**, 036002 (2014); Adv. High Energy Phys. **2014**, 305298 (2014).

[26] V.M. Lobashev *et al.*, Phys. Lett. B **460**, 227 (1999).

[27] V.N. Aseev *et al.*, Phys. Rev. D **84**, 112003 (2011).

[28] V.N. Aseev *et al.*, Phys. Atom. Nucl. **75**, 464 (2012).

[29] C. Kraus *et al.*, Eur. Phys. J. C **40**, 447 (2005).

[30] J. Angrik *et al.*, Preprint FZKA-7090 (2005).

[31] M. Aker *et al.*, Phys. Rev. Lett. **123**, 221802 (2019).
[32] M. Aker *et al.*, arXiv:2105.08533.
[33] A. Ashtari Esfahani *et al.* [Project 8], J. Phys. G **44**, 054004 (2017).
[34] R. Lehnert, Phys. Lett. B **828**, 137017 (2022).
[35] M. Aker *et al.*, arXiv:2207.06326.
[36] See, e.g., F. Šimkovic, R. Dvornický, and A. Faessler, Phys. Rev. C **77**, 055502 (2008).
[37] N. Kuroda *et al.*, Nature Commun. **5**, 3089 (2014).
[38] V.A. Kostelecký and A.J. Vargas, Phys. Rev. D **92**, 056002 (2015).
[39] C. Amole *et al.*, Nature **483**, 439 (2012).
[40] A. Kellerbauer *et al.*, Nucl. Instr. Meth. Phys. Res. B **266**, 351 (2008); S. Aghion *et al.*, Nature Commun. **5**, 4538 (2014); J. Storey *et al.*, Hyperfine Int. **228**, 151 (2014).
[41] P. Indelicato *et al.*, Hyperfine Int. **228**, 141 (2014); P. Perez and Y. Sacquin, Class. Quantum Grav. **29**, 184008 (2012); G. Chardin *et al.*, CERN report CERN-SPSC-2011-029, September 30, 2011.
[42] C. Amole *et al.*, Nat. Commun. **4**, 1785 (2013); H. Müller, P. Hamilton, A. Zhmoginov, F. Robicheaux, J. Fajans, and J. Wurtele, Phys. Rev. Lett. **112**, 121102 (2014).
[43] A.D. Cronin *et al.*, *Letter of Intent: Antimatter Antigravity Experiment (AGE) at Fermilab*, February 6, 2009.

EPJ Web of Conferences **282**, 01005 (2023) https://doi.org/10.1051/epjconf/202328201005
SSP 2022

Toward a pulsed antihydrogen beam for WEP tests in AEgIS

Saiva Huck[1,2,*], *Marcis* Auzins[3], *Benedikt* Bergmann[5], *Petr* Burian[5], *Roberto S.* Brusa[7,8], *Antoine* Camper[9], *Ruggero* Caravita[7,8], *Fabrizio* Castelli[10,11], *Roman* Ciuryło[12], *Giovanni* Consolati[10,13], *Michael* Doser[1], *Aaron* Farricker[19], *Lisa T.* Glöggler[1], *Łukasz* Graczykowski[14], *Malgorzata* Grosbart[1], *Francesco* Guatieri[7,8], *Natali* Gusakova[1,20], *Stefan* Haider[1], *Malgorzata* Janik[14], *Grzegorz* Kasprowicz[14], *Ghanshyambhai* Khatri[1,9], *Łukasz* Kłosowski[12], *Georgy* Kornakov[14], *Valts* Krumins[1,3], *Lidia* Lappo[14], *Adam* Linek[12], *Chloé* Malbrunot[1,15], *Sebastiano* Mariazzi[7,8], *Lilian* Nowak[1], *Dorota* Nowicka[14], *Emmanuel* Oswald[1], *Luca* Penasa[7,8], *Vojtech* Petracek[4], *Mariusz* Piwiński[12], *Stanislav* Pospisil[5], *Luca* Povolo[7,8], *Francesco* Prelz[10], *Sadiq* Rangwala[16], *Benjamin* Rienäcker[1,19], *Volodymyr* Rodin[19], *Ole M.* Røhne[9], *Heidi* Sandaker[9], *Petr* Smolyanskiy[5], *Tomasz* Sowiński[18], *Dariusz* Tefelski[14], *Marco* Volponi[1,7,8], *Carsten* Welsch[19], *Tim* Wolz[1,9], *Michal* Zawada[12], *Jakub* Zielinski[14], *Christian* Zimmer[9,13,1], and *Nicola* Zurlo[6,17]

[1]Physics Department, CERN, 1211 Geneva 23, Switzerland
[2]Institute for Experimental Physics, Universität Hamburg, Hamburg, Germany
[3]University of Latvia, Department of Physics Raina boulevard 19, LV-1586, Riga, Latvia
[4]Czech Technical University, Prague, Brehová 7, 11519 Prague 1, Czech Republic
[5]Institute of Experimental and Applied Physics, Czech Technical University in Prague, Czech Republic
[6]INFN Pavia, via Bassi 6, 27100 Pavia, Italy
[7]Department of Physics, University of Trento, via Sommarive 14, 38123 Povo, Trento, Italy
[8]TIFPA/INFN Trento, via Sommarive 14, 38123 Povo, Trento, Italy
[9]Department of Physics, University of Oslo, Sem Sælandsvei 24, 0371 Oslo, Norway
[10]INFN Milano, via Celoria 16, 20133 Milano, Italy
[11]Department of Physics "Aldo Pontremoli", University of Milano, via Celoria 16, 20133 Milano, Italy
[12]Institute of Physics, Nicolaus Copernicus University in Toruń, Grudziadzka 5, 87-100 Toruń, Poland
[13]Department of Aerospace Science and Technology, Politecnico di Milano, 20156 Milano, Italy
[14]Warsaw University of Technology, Faculty of Physics, ul. Koszykowa 75, 00-662, Warsaw, Poland
[15]Stefan Meyer Institute for Subatomic Physics, Austrian Academy of Sciences, 1090 Vienna, Austria
[16]Raman Research Institute, C. V. Raman Avenue, Sadashivanagar, Bangalore 560080, India
[17]Department of Civil, Environ., Archt. Eng. and Mathematics, University of Brescia, Brescia, Italy
[18]Institute of Physics, Polish Academy of Sciences, Aleja Lotnikow 32/46, PL-02668 Warsaw, Poland
[19]University of Liverpool, UK and The Cockcroft Institute, Daresbury, United Kingdom
[20]Department of Physics, Norwegian University of Science and Technology, Trondheim, Norway

Abstract. The AE$\bar{g}$IS collaboration at CERN's AD produces antihydrogen atoms in the form of a pulsed, isotropic source with a precisely defined formation time. AE$\bar{g}$IS has recently undergone major upgrades to fully benefit from the increased number of colder antiprotons provided by the new ELENA decelerator and to move toward forming a horizontal beam to directly investigate the influence of gravity on the $\bar{H}$ atoms, thereby probing the Weak Equivalence Principle for antimatter. This contribution gives an overview of these upgrades as well as subsequent results from the first beam times with ELENA.

*e-mail: saiva.huck@cern.ch

1 Introduction

The AE$\bar{g}$IS (Antimatter Experiment: Gravity, Interferometry, Spectroscopy) collaboration at CERN's Antiproton Decelerator (AD) complex aims at probing the Weak Equivalence Principle (WEP) via a direct measurement of the gravitational acceleration of antimatter by means of a horizontal beam of antihydrogen atoms.
AE$\bar{g}$IS forms Rydberg antihydrogen inside a Penning-Malmberg trap via a charge exchange reaction between cold antiprotons and positronium atoms laser-excited to Rydberg states:

$$Ps^{*} + \bar{p} \rightarrow \bar{H}^{*} + e^{-}. \tag{1}$$

The long-term aim is a measurement of the influence of gravity on these $\overline{\mathrm{H}}$ atoms, whose production time is known with a precision of a few hundred nanoseconds by forming a horizontal beam that traverses a known distance through a deflectometer to a position-sensitive detector, registering their vertical displacement [1].
During the AD run in 2018, individual $\overline{\mathrm{H}}$ atoms were successfully produced in the AE$\bar{g}$IS apparatus in an isotropic, pulsed scheme [2]. This concluded the initial phase of the experiment, having demonstrated the feasibility of the approach for $\overline{\mathrm{H}}$ production with a well-defined formation time. AE$\bar{g}$IS is now entering the subsequent stage: improving conditions to increase the rate of cold $\overline{\mathrm{H}}$ production, enabling efficient beam formation and gravity measurements.
With the installation of an additional decelerator, ELENA (Extra Low ENergy Antiproton ring), in the AD complex, the antiprotons directly available to the experiments have a strongly reduced energy of 100 keV, significantly improving the possible capture efficiency. Furthermore, ELENA enables 24 hour $\overline{\mathrm{p}}$ provision. AE$\bar{g}$IS has taken the occasion of reconstructing the entire $\overline{\mathrm{p}}$ entrance region to connect to ELENA as well as to improve operation for the first beam time in fall of 2021 to also upgrade several further components essential to fully profit from ELENA and move toward beam formation.

2 AE$\bar{g}$IS Phase 2

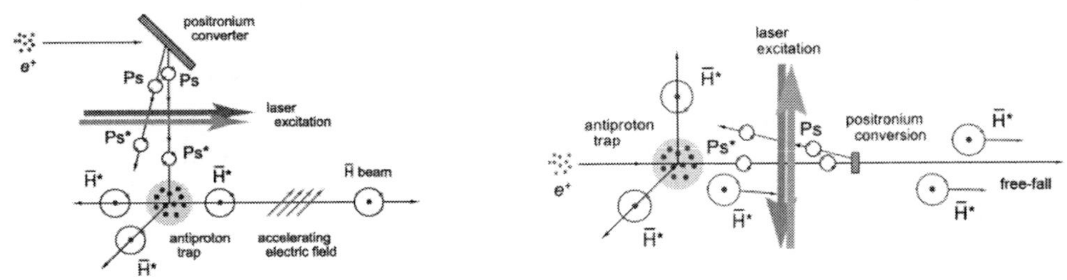

Figure 1: Schematics of antihydrogen production in AE$\bar{g}$IS. Left: orthogonal scheme used for AE$\bar{g}$IS Phase 1. Right: collinear scheme used for AE$\bar{g}$IS Phase 2.

To reduce solid angle and ionization losses of Ps, the AE$\bar{g}$IS $\overline{\mathrm{H}}$ production has been completely re-organized. For Phase 1, antihydrogen was produced by inserting Ps atoms through a hole in the top of the trap confining the antiproton plasma, as shown in fig. 1, left. Since the Ps atoms are excited to Rydberg states to increase their lifetime and benefit from the scaling of the charge exchange cross section [3], they experienced dynamical field ionization as they moved orthogonally in the magnetic field to reach the trap. The resulting losses limited the excitation to $n_{Ps} = 17$. To mitigate this effect for Phase 2, $\overline{\mathrm{H}}$ production has been transformed into the collinear setup depicted on the right side of fig. 1, with Ps produced on-axis.

2.1 Hardware upgrades

The adaptation to a collinear scheme has the added benefit of eliminating the need for an opening in the top of the $\overline{\mathrm{H}}$ production trap. Consequently, a brand new Penning-Malmberg trap without azimuthal asymmetries and with larger electrode radii has been constructed to prevent expansion and heating of the plasma (fig. 2, left photo). In addition to an improved monitoring for Ps excitation, the new trap incorporates further upgrades: 1. a set of cryogenic actuators to enable accurate alignment of the trap with respect to the magnetic axis and movement of the $\overline{\mathrm{H}}$ ionization grid at operating conditions, and 2. a novel e^+ to Ps conversion target installed on-axis on another such actuator and allowing for a deeper implantation of e^+ into the target to produce lower velocity Ps. The target consists of a $5\times5\times1\ \mathrm{mm}^3$ Si nanostructure on a 3D-printed Ti holder with included temperature control (fig. 2, central left image). Minimizing the dimensions of the whole structure is a requirement of the collinear $\overline{\mathrm{H}}$ production to limit solid angle losses as the $\overline{\mathrm{H}}$ beam travels axially outwards past the target.
Further hardware upgrades to the AE$\bar{\mathrm{g}}$IS apparatus include an entirely new laser system and transfer line for a more efficient Ps excitation, modifications to the central scintillating fiber tracking detector to allow for a larger sensitive area, and a new degrader setup. Thanks to the lower $\bar{\mathrm{p}}$ energies from ELENA, much thinner degrader foils are needed to achieve the final reduction to a trappable energy ($< 10\,\mathrm{keV}$). AE$\bar{\mathrm{g}}$IS Phase 2 is using Mylar and Parylene N foils with thicknesses of several hundreds of nanometers. The first degradation step is accomplished by the combination of a 1400 nm foil with a four-fold segmented, 10 nm aluminium layer and a coating-free central cross, which can thus simultaneously act as a rough beam position monitor, plus an additional, fully Al-coated, 100 nm foil that can also be used as a Faraday Cup. The signals of both can be obtained directly and via cryogenic amplifiers mounted on a circular PCB (fig. 2, central right photo). To perform the final degradation fine tuning, additional foils with thicknesses between 100 nm and 500 nm have been installed on an actuator-controlled ladder (fig. 2, right image).

Figure 2: Photographs of upgraded hardware components in AE$\bar{\mathrm{g}}$IS. Left: $\overline{\mathrm{H}}$ production trap. Central left: positronium target on holder structure. Central right: main degrader with segmented Al coating. Right: Foil ladder for degradation fine tuning.

2.2 The new control system

One of the most central upgrades of AE$\bar{\mathrm{g}}$IS concerns the implementation of a completely new experimental control system, necessary for a full automation of the control and synchronization of all involved hardware as well as the data taking procedures and to render the experiment maintainable and adaptable in the long term.
The electronics of the new control system are based on Sinara hardware [4], which is designed for dedicated control and nanosecond synchronization using the ARTIQ control software [5]. The main control carrier is an Artix-7 FPGA, which, in the case of AE$\bar{\mathrm{g}}$IS, controls digital I/O units, fast 16-bit DAC modules, and custom-designed 1 MHz 200-fold high-voltage amplifiers for each DAC channel. Several such complete units have been commissioned and calibrated to precisely control the trap electrode potentials as well as to synchronize all involved sub-systems of the experiment.

From the software point of view, AE$\bar{g}$IS Phase 2 is being controlled by a newly constructed LabVIEW^{TM} control system, which manages the communication between all integral parts of the experiment, enables 24-hour automatized operation, and feeds the defined run routines to the hardware controllers. The LabVIEW^{TM} system is based on the division of the code into dedicated, autonomous components, which are monitored and synchronized by a distributed system of guardian processes. This approach provides the system with strong stability and facilitates a highly efficient use of beam time.

3 First beam times with ELENA

AE$\bar{g}$IS used the first ELENA $\bar{p}$ runs for in-depth tests of large parts of the upgraded hard- and software with a particular focus on the stability and operation of the control system.
In summary, the accurate timing control of the new system was verified, making full use of the precise triggering and coordinating the involved sub-systems accordingly, and close to no round-the-clock beam time was lost thanks to the full automation and stable operation. As shown in fig. 3, AE$\bar{g}$IS has furthermore successfully demonstrated antiproton degradation, capture and manipulation with the new system, paving the way for an improved antihydrogen production in the subsequent runs.

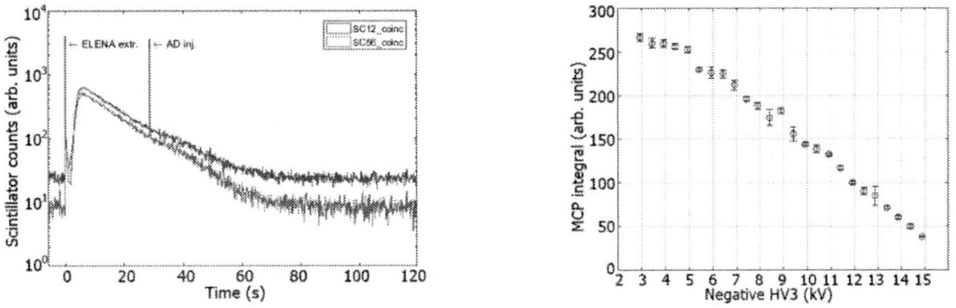

Figure 3: First results of $\bar{p}$ beam time with ELENA in AE$\bar{g}$IS. Left: $\bar{p}$ capture and annihilation on rest gas, monitored via two sets of external scintillators. Right: Amount of $\bar{p}$ passing the barrier of an HV electrode after energy degradation, as observed on a downstream MCP.

4 Conclusions

Having demonstrated the successful operation of a majority of the required hardware and software upgrades, AE$\bar{g}$IS is moving toward its main goal of efficiently producing cold antihydrogen and forming a horizontal $\bar{H}$ beam.
Once formed, the beam will traverse the grids of a deflectometer and its gravity-induced displacement will be measured on a time- and position-sensitive detector, while the pulsed production scheme yields precise knowledge of the passage time. Combining the data for both parameters, it is then possible to draw direct conclusions on the gravitational acceleration. Such a measurement will be a first step toward probing the WEP for antimatter.

References

[1] M. Doser et al., Class. Quantum Grav. **29**, 184009 (2012)
[2] C. Amsler et al., Nature Comms. Phys. **4**:19 (2021)
[3] D. Krasnicky et al., Phys. Rev. A **94**, 022714 (2016)
[4] G. Kasprowicz et al., OSA Quantum 2.0 Conference, QTu8B.14 (2020)
[5] S. Bourdeauducq et al., Zenodo, https://doi.org/10.5281/zenodo.51303 (2016)

EPJ Web of Conferences **282**, 01006 (2023) https://doi.org/10.1051/epjconf/202328201006
SSP 2022

Origin of the Proton Mass

Craig D. Roberts[1,2,*]

[1]School of Physics, Nanjing University, Nanjing, Jiangsu 210093, China
[2]Institute for Nonperturbative Physics, Nanjing University, Nanjing, Jiangsu 210093, China

Abstract. Atomic nuclei lie at the core of everything visible; and at the first level of approximation, their atomic weights are simply the sum of the masses of all the neutrons and protons (nucleons) they contain. Each nucleon has a mass $m_N \approx 1$ GeV $\approx$ 2000-times the electron mass. The Higgs boson – discovered at the large hadron collider in 2012, a decade ago – produces the latter, but what generates the nucleon mass? This is a pivotal question. The answer is widely supposed to lie within quantum chromodynamics (QCD), the strong-interaction piece of the Standard Model. Yet, it is far from obvious. In fact, removing Higgs-boson couplings into QCD, one arrives at a scale invariant theory, which, classically, can't support any masses at all. This contribution sketches forty years of developments in QCD, which suggest a solution to the puzzle, and highlight some of the experiments that can validate the picture.

1 Introduction

The 2013 Nobel Prize in Physics was awarded to Englert and Higgs following discovery of the Higgs boson at the Large Hadron Collider (LHC) [1, 2]. With this discovery, the Standard Model of Particle Physics (SM) became complete; and some began to ask: "Where do we go from here?" The beginning of an answer lies in recognising just what the SM is.

As formulated, the SM offers a description of all known fundamental physics except gravity; and since gravity has no discernible effect when particles are studied a few at a time, one can largely neglect it when considering subatomic physics [3]. Since it discovery at the LHC [4, 5], the Higgs Boson has been promoted to the *Centre of Things* – see, *e.g.*, Ref. [6]. In the lead image of that article it is apparent that the SM has 17 particles. It also has 19 parameters, most of which relate to the Higgs and all of which must be determined through comparison with experiment. So, although the SM supposedly describes the most powerful forces in Nature, with so many free parameters, it is somewhat unsatisfactory.

Many will have heard one or another variant of the oft made remark: "All particles gain their mass from a fundamental field associated with the Higgs boson." There is some truth in this. The Higgs does seem to be the source of the mass of elementary particles, *e.g.*, the electron; but it is responsible for < 2% of the mass of more complex things, like the proton. The mass of the vast bulk of visible material in the Universe has a different source.

Strong interactions in the SM are expected to be explained by quantum chromodynamics (QCD) [7]. The Higgs reaches into QCD: it is the source of the Lagrangian current masses of the six known quarks. These are 6 of the SM's 19 parameters. However, only 2 of those

*e-mail: cdroberts@nju.edu.cn

19 parameters are intrinsic to QCD. One of them is $\theta_{\rm QCD}$, which appears to be (almost) zero because the nucleon electric dipole moment is, as yet, unmeasurably small [8]. This leaves just one parameter in QCD that needs to be fixed – a mass-scale, typically denoted $\Lambda_{\rm QCD}$. So, perhaps science has a chance of understanding at least this one piece of the SM?

Or, perhaps not. In many ways, QCD is the most important piece of the SM because it is supposed to describe all nuclear physics. It is formulated in terms of matter fields – six current quarks – and gauge bosons – eight gluons. Yet, fifty years after the discovery of quarks [9–12], science is only just beginning to understand how QCD moulds the basic elements of nuclei, *viz.* pions, neutrons, protons, *etc.* Moreover, there are controversies, as theory begins to predict quantities that hitherto were only inferred from measurements via phenomenological fits.

To proceed, let's assume that quantum gauge field theories are the right paradigm for understanding Nature. It is then important that we find ourselves in a Universe where time and space give us four (noncompact) dimensions. $D = 4$ is a critical point because quantum field theories in $D \neq 4$ dimensions possess an explicit mass-scale: their couplings carry a mass dimension. Theories in $D > 4$ are characterised by uncontrollable ultraviolet divergences. On the other hand, those in $D < 4$, while superconvergent, are afflicted with a hierarchy problem: dynamical effects typically contribute $\lesssim 10\%$ to any mass-dimensioned quantity [13–17].

In contrast, omitting Higgs couplings, the $D = 4$ SM is built from scale-invariant classical field theories. Each is renormalisable, a process that introduces a mass scale [18, Ch. III]. The size of that mass is not determined by the theory. So, what determines the natural scale for visible matter? We know it is $m_{\rm Nature} \approx m_{\rm proton} \approx 1\,$GeV; but how much "wiggle room" is there in that number, *i.e.*, what is the window of values for $\delta m_{\rm Nature}$ such that $m_{\rm Nature} \to m_{\rm Nature} + \delta m_{\rm Nature}$ would still deliver a habitable Universe? Perhaps an ultimate theory of Nature will answer these questions, but no one can be sure that such a theory exists.

This brings us to the central points. Namely, the existence of our Universe depends critically on, *inter alia*, the following empirical facts. (*a*) The proton is massive, by which one means that the mass-scale for strong interactions is much larger than that of electromagnetism. (*b*) It is also absolutely stable, despite being a composite object constituted from three light valence quarks. (Here, light means that the proton's valence quarks have masses not much more than that of the electron [19].) (*c*) On the other hand, the π-meson is unnaturally light: its mass is commensurate with that of the μ-lepton, despite the π being a strongly interacting composite object built from a valence-quark and valence antiquark. These things and their consequences are, supposedly, written in the QCD Lagrangian; but they are far from obvious. Indeed, they are examples of emergence [20–26], *viz.* low-level rules producing high-level phenomena, with enormous apparent complexity.

2 Emergence of Hadron Mass

It is now plain that the SM has one known source of mass, *i.e.*, the Higgs boson (HB). Its contributions are critical to the evolution of our Universe. However, as shown in Fig. 1A, the Higgs is responsible for only 9 MeV in the 939 MeV of proton mass; hence, just 1% of the mass of the visible material in the Universe. Plainly, Nature has another, very effective mechanism for producing mass, which has come to be known as emergent hadron mass (EHM). Alone, EHM is responsible for 94% of $m_{\rm proton}$. The remaining 5% arises from constructive interference between these two sources.

At this point, many will be asking: "But what is the origin of EHM?" It's a key question; many others are related. For instance: How is EHM connected with gluon and quark confinement – part of the \$1-million prize Millennium Problem [27]; with dynamical chiral symmetry breaking – the foundation stone for the existence of Nambu-Goldstone bosons; and

 https://doi.org/10.1051/epjconf/202328201006

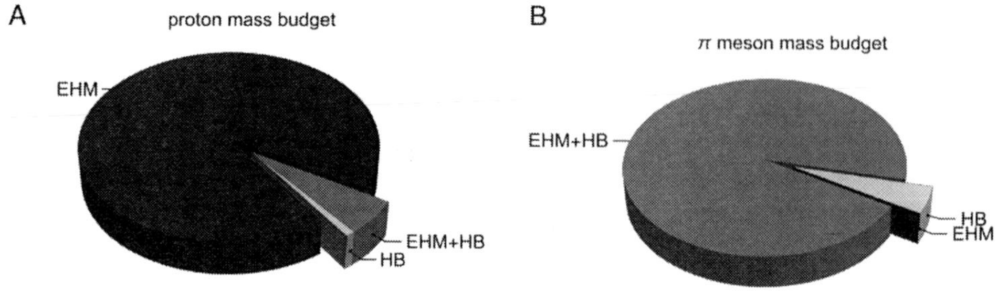

Figure 1. Mass budgets: A – proton; B – pion. Each is drawn using a Poincaré invariant decomposition, with the numerical values listed elsewhere [26, Table 1]: HB – contribution owing solely to the Higgs boson; EHM – contribution from emergent hadron mass; EHM+HB – mass generated by constructive interference between these two sources of mass.

what is the HB's role in modulating the observable properties of matter? This last issue is critical because, *e.g.*, without the Higgs, all current-quarks would be massless and the π- and K-meson would be indistinguishable. So take a look at the π-meson mass budget in Fig. 1B. There is no EHM-only contribution in this case because, being a Nambu-Goldstone boson, the pion is massless in the chiral limit, *i.e.*, with HB couplings removed from QCD. Even using realistic couplings, the HB part is still small – just 5%. But EHM has a prodigious indirect impact: EHM+HB interference is responsible for 95% of m_π [20, Sec. V].

In defining QCD, one begins with a one-line Lagrangian density:

$$\mathcal{L}_{\rm QCD} = \sum_{f=u,d,s,\ldots} \bar{q}_f[\gamma\cdot\partial + ig\tfrac{1}{2}\lambda^a\gamma\cdot A^a + m_f]q_f + \tfrac{1}{4}G^a_{\mu\nu}G^a_{\mu\nu}, \tag{1}$$

where $\{q_f\,|\,f = u, d, s, c, b, t\}$ are fields associated with the six known flavours of quarks; $\{m_f\}$ are their current-masses, generated by the Higgs boson; $\{A^a_\mu\,|\,a = 1,\ldots,8\}$ represent the gluon fields, whose matrix structure is encoded in $\{\frac{1}{2}\lambda^a\}$, the generators of SU(3) in the fundamental representation; $G^a_{\mu\nu} = \partial_\mu A^a_\nu + \partial_\nu A^a_\mu - gf^{abc}A^b_\mu A^c_\nu$ is the gluon field strength tensor; and g is the *unique* QCD coupling, using which one conventionally defines $\alpha = g^2/[4\pi]$. A peculiar feature of Eq. (1) is that this Lagrangian density is expressed in terms of gluon and quark parton fields, which are *not* the degrees-of-freedom measured in detectors. This raises new questions. Namely, what are the asymptotic, detectable degrees-of-freedom; and how are they built from the Lagrangian partons? If such questions can be answered, then we might learn whether QCD is the theory of strong interactions; indeed, whether it is really a theory. These things are crucial, because, hitherto, science has not delivered a mathematically well defined four dimensional quantum field theory.

The only mass scales manifest in Eq. (1) are the current-quark masses, $\{m_f\}$. Insofar as the proton is concerned, only up (u) and down (d) quarks are immediately relevant because the proton is a bound-state of the valence-quark combination $u + u + d$. As apparent from Fig. 1, the current mass of any one of these quarks is just $1/300^{\rm th}$ of $m_{\rm Nature}$, *viz.* more than two orders-of-magnitude smaller. Plainly, the HB generated mass is very far removed from the natural scale for strongly-interacting matter. Moreover, no amount of staring at $\mathcal{L}_{\rm QCD}$ can reveal that scale. This is in stark contrast to quantum electrodynamics (QED), in which, *e.g.*, the electron mass that appears explicitly in $\mathcal{L}_{\rm QED}$ provides the mass unit for the spectrum of hydrogen energy levels.

So, let's remove the current-quark masses, *i.e.*, suppress HB couplings into QCD. Then no explicit scales remain and the action associated with Eq. (1) is scale invariant. There is no dynamics in a scale invariant theory; only kinematics survive, and the theory appears the same at all length scales. Thus, isolated clumps of material can't form, which means bound states are impossible. Consequently, our Universe can't exist. As noted above, the Higgs doesn't solve this problem because u, d quarks are *light* and the masses of protons and neutrons – the kernels of all visible matter – are > 100-times larger than anything the Higgs produces in this quark sector. Thus, the question of how our Universe came to be is inseparable from that of "How do nucleons become so massive?" (A detailed discussion of this and the following points in this section may be found in Ref. [28].)

All theories have an energy-momentum tensor, $T_{\mu\nu}$ – energy and momentum conservation are expressed in the divergence relation $\partial_\mu T_{\mu\nu} = 0$. The Noether current associated with spacetime dilations is $\mathcal{D}_\mu = T_{\mu\nu}x_\nu$; and in a scale invariant theory, $\partial_\mu \mathcal{D}_\mu = T_{\mu\mu} \equiv 0$. In quantising chromodynamics, a mass scale is introduced by the process of regularisation and renormalisation of ultraviolet divergences. This entails that the constants in $\mathcal{L}_{\rm QCD}$, *i.e.*, coupling and current masses, come to depend on the renormalisation scale, hereafter denoted ζ. Let's switch off the HB couplings again, then one need only consider the QCD fine structure constant: $\alpha \to \alpha(\zeta)$. In this case, some straightforward algebra involving Eq. (1) reveals that, under the change of scale $\zeta \to e^\sigma \zeta$, $\alpha \to \sigma\alpha\beta(\alpha)$, where $\beta(\alpha)$ is the QCD "β-function", and

$$\mathcal{L}_{\rm QCD} \to \sigma\,\alpha\beta(\alpha)\,\frac{\delta \mathcal{L}_{\rm QCD}}{\delta\alpha} \Rightarrow \partial_\mu \mathcal{D}_\mu = \frac{\delta \mathcal{L}_{\rm QCD}}{\delta\sigma} = \alpha\beta(\alpha)\,\frac{\mathcal{L}_{\rm QCD}}{\delta\alpha} = \beta(\alpha)\tfrac{1}{4}G^a_{\mu\nu}G^a_{\mu\nu} = T_{\mu\mu} =: \Theta_0\,. \tag{2}$$

Evidently, quantisation of our four-dimensional theory has introduced a nonzero value for the trace of energy-momentum tensor. This is the (in)famous QCD trace anomaly.

The existence of a trace anomaly means that a mass scale must exist in QCD. The key issue is whether, or not, one can compute that scale and/or understand its magnitude. Prediction of the scale's value is beyond the SM. On the other hand, measurement is straightforward. Consider a proton with four-momentum P, represented by a state vector $|p(P)\rangle$, then in a Poincaré invariant theory:

$$\langle p(P)|T_{\mu\nu}|p(P)\rangle = -P_\mu P_\nu \Rightarrow \langle p(P)|T_{\mu\mu}|p(P)\rangle = -P_\mu P_\mu = m^2_{\rm proton} = \langle p(P)|\Theta_0|p(P)\rangle. \tag{3}$$

Thus, absent HB couplings, the entirety of the proton mass owes to the trace anomaly. Looking at Eq. (2), then from this perspective, Θ_0 measures the in-proton strength of gluon self-interactions; so, $m_{\rm proton} \approx 1$ GeV is somehow generated completely by $m = 0$ gluon partons.

Let's consider the other mass budget drawn above, *viz*. that for the π-meson in Fig. 1B. In the absence of HB couplings, $m_\pi = 0$; hence,

$$\langle \pi(q)|T_{\mu\nu}|\pi(q)\rangle = -q_\mu q_\nu \Rightarrow \langle \pi(q)|T_{\mu\mu}|\pi(q)\rangle = -q_\mu q_\mu = m^2_\pi = 0 = \langle \pi(q)|\Theta_0|\pi(q)\rangle. \tag{4}$$

Does this mean that the scale anomaly vanishes trivially in the pion state; to wit, gluons contribute nothing to the pion mass? Some have claimed this – see Ref. [29, Sec. 2]; but given that all the proton's mass owes to Θ_0, it seems unlikely. Far more plausibly, "0" owes to cancellations between different operator contributions to the in-π expectation value of Θ_0. In physics, of course, such precise cancellations are never accidental. Typically, they only arise because some symmetry is dynamically broken – see Eq. (6) below.

It is now abundantly clear that one cannot address the question of the origin of $m_{\rm proton}$ without simultaneously explaining how the π-meson remains massless. The natural visible-matter mass-scale must emerge coextensively with apparent preservation of scale invariance in related systems. Moreover, the in-pion expectation value of Θ_0 is always zero, irrespective

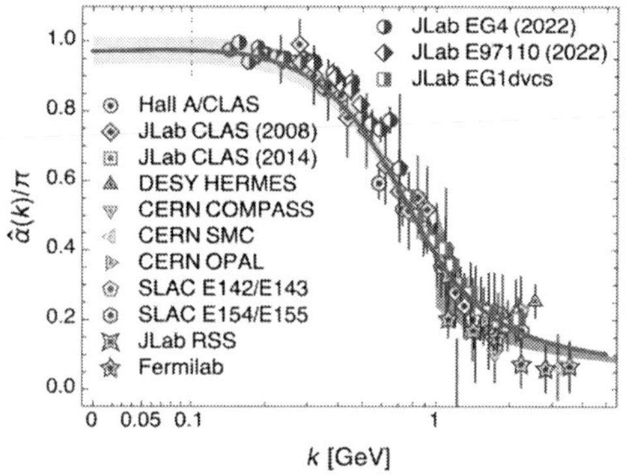

Figure 2. Process-independent effective charge, $\hat{\alpha}(k)/\pi$, obtained by combining modern results from continuum and lattice analyses of QCD's gauge sector [38]. Note the dramatic change in behaviour on $k \lesssim m_0$, as perturbative interactions are overwhelmed by EHM. Existing data on the process-dependent charge, α_{g_1}, are also drawn. Their sources are discussed in Ref. [39] and the meaning of the comparison in Refs. [23, 26]. (Image courtesy of D. Binosi.)

of the size of m_{proton}. The challenge to modern physics is to elucidate the manifold empirical consequences of the mechanism responsible so that the SM can be validated. (This does not overlook the fact that no finite number of measurements can validate a theory. Instead, it asks for SM boundaries to be pushed in new directions within strong interaction physics with a view to testing whether $D = 4$ quantum non-Abelian gauge field theories might be a viable paradigm for extending the SM.)

3 Three Pillars of EHM

The path to understanding EHM was laid sixty years ago, with the first demonstration that gauge bosons could dynamically acquire mass without destroying any of the essential features of gauge theories [30, 31]. Forty years ago, it was realised that this "Schwinger mechanism" of gauge boson mass generation could be active in QCD [32]. Since then, the idea has been refined [33–37], so that a detailed picture is now emerging. Continuum theory and numerical simulations of lattice-regularised QCD both agree that owing to gluon self-interactions, gluon partons are transformed into gluon quasiparticles, which are characterised by a momentum dependent mass function that is large at infrared momenta. The associated renormalisation group invariant scale is [38]:

$$m_0 = 0.43(1)\,\text{GeV} \approx \tfrac{1}{2} m_{\text{Nature}}. \tag{5}$$

Here one truly sees *mass emerge from nothing*: an interacting $D = 4$ theory, expressed using massless gluon partons, dynamically generates massive dressed gluon fields. This is the first pillar of EHM. Now let's consider its consequences.

A crucial step on the road to presenting QCD as the theory of strong interactions was the discovery of asymptotic freedom [3, 40, 41], [42, Ch. 7.1]: the interaction between charge-carriers becomes weaker as the momentum-squared characterising the scattering process, k^2, becomes larger. (QED exhibits the opposite behaviour [43, Ch. 13].) The perturbative analysis of this coupling delivered a result that diverges at $k^2 = \Lambda^2_{\text{QCD}} < m^2_{\text{proton}}$. This is the "Landau pole", which can't be eliminated in perturbation theory. On the other hand, modern methods in theory have enabled calculation of the unique QCD analogue of QED's Gell-Mann–Low effective charge, which extends the perturbative form onto the strong coupling domain. The solution is drawn in Fig. 2. It is a parameter-free QCD prediction. Physically, the generation of a gluon mass scale has ensured that long wavelength gluons are screened from interactions, eliminating the Landau pole, so that the QCD running coupling saturates to a finite value in the far infrared. This running coupling is the second pillar of EHM.

The emergence of mass in QCD's gauge sector is communicated to the matter sector, with the appearance of a chiral-limit running mass for quarks, $M_0(k)$, that reaches a value at infrared momenta whose size is commensurate with the constituent-quark mass: $M_0(0) \lesssim m_0$. Illustrated and discussed elsewhere [23, Fig. 2.5], this is the third pillar of EHM.

These three pillars of EHM are expressed in every strong interaction observable, even if that is only by giving a new stability to perturbation theory through the emergence of dynamically generated infrared cutoffs. The theory challenge is to elucidate their diverse measurable expressions; and the experimental challenge is to test the predictions and establish the soundness of the paradigm: simplicity to elegant complexity in the emergence of the most fundamental structures in Nature.

4 EHM at Existing and Future Facilities

To meet the theory and experimental challenges, science must move beyond the 100-year focus on proton structure. Columbus did not discover the New World by meticulously charting the lagoon at *Palos de la Frontera*; and QED was not revealed by careful studies of the hydrogen ground state alone. Contemporary theory, which connects the three pillars of EHM to hadron observables, reveals extraordinary variations in structure between kindred and neighbouring states; so, science must use every tool at its disposal and develop new facilities that can expose the hand of EHM in the spectrum of all hadrons and their structure. The theory predictions and experimental opportunities are canvassed in numerous sources [20–26, 29, 44–50]. Herein, owing to limitations of space and time, I will only touch upon EHM as it is relates to future measurements of proton and pion structure functions.

Considering Eq. (4), it is clear that the pion is special. The fact that $\langle\pi(q)|\Theta_0|\pi(q)\rangle \equiv 0$ in the absence of HB couplings is explained by a simple yet potent QCD identity; namely [51]

$$\begin{array}{ccc} \text{two body problem} & & \text{one body problem} \\ f_\pi^0 E_\pi^0(k;P) & = & B_0(k^2) \end{array} . \tag{6}$$

The left-hand side of this identity contains a product of terms that measures the leading component in the pion's bound-state amplitude; the right-hand side is the scalar piece of the dressed-quark self-energy; and this identity states that the pseudoscalar meson two-body problem is solved, almost completely, once the solution of the dressed-quark one body problem is known. (The "0" super/subscript means HB couplings turned off.) Eq. (6) is a corollary of EHM: it is a sufficient and necessary condition for dynamical chiral symmetry breaking; provides the QCD foundation for the success of chiral effective field theory; and means that the properties of the (nearly) massless pion are the cleanest expressions of EHM in Nature.

Viewed from a perspective defined by $\mathcal{L}_{\rm QCD}$ in Eq. 1, protons and pions appear as sketched schematically in Fig. 3. This structure is characterised by light-front momentum-fraction distribution functions (DFs) of valence quark partons, sea quark partons, and gluon partons, which can, in principle, be extracted from analyses of structure function measurements. However, despite enormous expense of time and effort, much must still be learnt before the proton and pion may be considered understood in terms of these DFs. Most simply, what are the differences, if any, between the distributions of partons within the proton and the pion? Owing to a growing appreciation of the need to understand EHM, this question now has particular resonance as science attempts to determine how the obvious macroscopic differences between protons and pions are expressed in their DFs.

Regarding measurements that do not involve beam or target polarisation, it is known that there is a scale – the hadron scale, $\zeta_{\mathcal{H}} < m_{\rm proton}$, at which valence-quark DFs in the proton

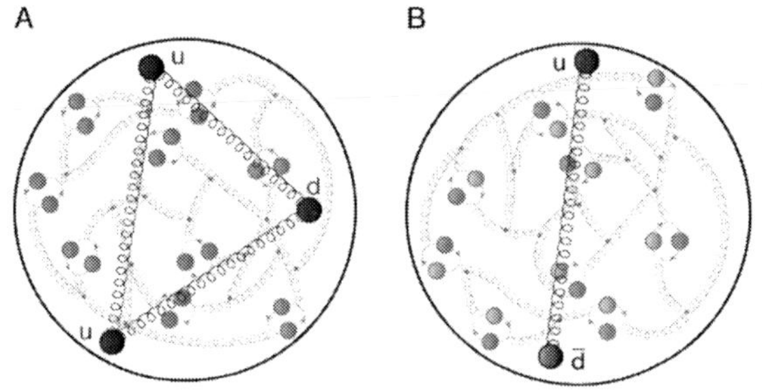

Figure 3. A. In terms of QCD's Lagrangian quanta, the proton, p, contains two valence up (u) quarks and one valence down (d) quark; and also infinitely many gluons and sea quarks, drawn here as "springs" and closed loops, respectively. B. The pion, π^+, contains one valence u-quark, one valence $\bar{d}$-quark, and, akin to the proton, infinitely many gluons and sea quarks.

and pion behave as follows [52–56]:

$$d^p(x;\zeta_{\mathcal{H}}), u^p(x;\zeta_{\mathcal{H}}) \overset{x\simeq 1}{\propto} (1-x)^{\beta^p_{\mathcal{H}}=3}\,, \quad \bar{d}^\pi(x;\zeta_{\mathcal{H}}), u^\pi(x;\zeta_{\mathcal{H}}) \overset{x\simeq 1}{\propto} (1-x)^{\beta^\pi_{\mathcal{H}}=2}\,. \qquad (7)$$

The light-front momentum fraction is bounded: $0 \leq x \leq 1$. It subsequently follows from the QCD evolution equations [57–60, DGLAP] that the large-x power on the related glue DF is $\approx \beta^{p,\pi}_{\mathcal{H}} + 1$ and that for sea quark DFs is $\approx \beta^{p,\pi}_{\mathcal{H}} + 2$. Moreover, as the scale increases to $\zeta > \zeta_{\mathcal{H}}$, these exponents all grow logarithmically. However, such QCD constraints are typically ignored in fits to the world's data on structure functions and kindred observables [61–65]. This is fueling controversy, leading some to question the veracity of QCD [54, 55, 66]. In addition, largely because pion data are scarce [23, Table 9.5], proton and pion data have never been considered simultaneously.

In this context, the unified array of results in Ref. [67], which uses a symmetry-preserving continuum Schwinger function method (CSM) to simultaneously predict the pointwise behaviour of all proton and pion DFs – valence, glue, and four-flavour-separated sea – is of significant value. Fig. 4 displays predictions for valence-quark and glue DFs at the scale $\zeta = m_{J/\psi} = 3.1$ GeV. The following features are worth highlighting. (*i*) At $\zeta = \zeta_{\mathcal{H}}$, valence quark degrees-of-freedom carry all the hadron's momentum: $\langle x\rangle^{\zeta_{\mathcal{H}}}_{u_p} = 0.69$; $\langle x\rangle^{\zeta_{\mathcal{H}}}_{d_p} = 0.31$; $\langle x\rangle^{\zeta_{\mathcal{H}}}_{\bar{d}_\pi} = \langle x\rangle^{\zeta_{\mathcal{H}}}_{u_\pi} = 0.5$. This is the defining characteristic of $\zeta_{\mathcal{H}}$. (*ii*) Quark+quark (diquark) correlations within the proton, whose formation is driven by EHM, ensure $u^p(x;\zeta) \neq 2d^p(x;\zeta)$. The predicted result for the ratio of these two functions agrees with modern experiments [68–70]. Indeed, the data reveal that the proton must contain both scalar and axialvector diquarks: the probability that the data can be explained by a scalar-diquark-only proton wave function is $1/141,000$ [26, Fig. 27]. (*iii*) Proton and pion DFs have markedly different behaviour on $x > 0.1$, which is called the valence-quark domain, with all pion DFs being markedly more dilated. In fact, pion DFs are the most dilated in Nature. This is a dramatic difference between proton and pion DFs. It follows from Eq. (6) and is a "smoking gun" signal for EHM.

It should also be stressed that the CSM predictions comply with QCD constraints on DF endpoint (low- and high-x) scaling behaviour, *e.g.*, Eq. (7). This is in stark contrast to existing global fits to data, which ignore them; so, fail to deliver realistic DFs, even from abundant proton data, and return controversial fits to the limited pion data. One may reasonably argue that only after imposing QCD constraints on future phenomenological data fits will it be possible to draw reliable pictures of hadron structure. This is especially important for attempts to empirically expose structural differences between Nambu-Goldstone bosons and seemingly less complex hadrons.

EHM is expressed in numerous other observables. For example, theory predicts that many excited states of the proton also contain pseudoscalar and vector diquark correlations. This can be tested in resonance electroexcitation measurements [71–73]. In addition, electroweak transitions between heavy+light mesons, in which HB couplings are the dominant source of

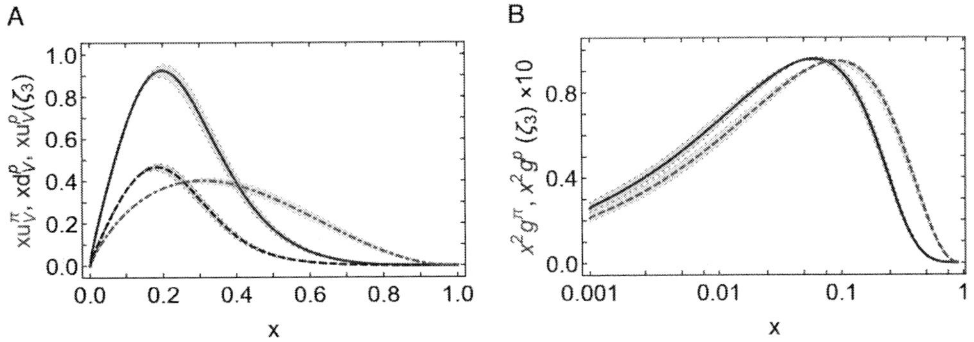

Figure 4. Distribution functions. A – valence quark DFs in the proton (u quark, solid red; d-quark, dashed blue) and pion (dashed green); B – glue DFs in the proton (solid purple) and pion (dashed green). In each case, the like coloured band expresses the response to a ±5% variation in $\zeta_{\mathcal{H}}$.

mass, and light (lighter) final states, whose mass is far more reliant on EHM, provide ideal opportunities to measure the impact of interference between Nature's two mass-generating mechanisms [26, Sec. 10]. In all these areas, progress demands a synergy between experiment, phenomenology, and theory.

5 Conclusion

QCD is an exceptional fundamental theory of natural phenomena. Its Lagrangian degrees-of-freedom are not directly observable. The Lagrangian's massless gluon partons become massive without "human intervention"; and their mass guarantees the existence of a unique running coupling that is everywhere finite, ensuring a stable infrared completion of the theory. Even in the absence of Higgs-boson coupling, the massless quarks become massive and interact to produce both massive hadrons and massless Nambu-Goldstone bosons. These effects form the three pillars of emergence in QCD. They are expressed in every strong interaction observable and can also be revealed via studies that contrast systems in which HB couplings provide the largest source of mass and those in which emergent mass is dominant.

Today, science is capable of building facilities that can validate these concepts, and thereby, possibly, elevate QCD, from its place as just one more effective field theory in the Standard Model, to a new position, as the first well-defined four-dimensional quantum field theory we've ever contemplated. That step may open doors that lead far beyond the Standard Model because, while we have theories of many things, we are still asking whether there is a theory of everything.

Acknowledgments. This contribution is based on insights developed through collaborations with many people, to all of whom I am greatly indebted. Work supported by: National Natural Science Foundation of China (grant no. 12135007).

References

[1] F. Englert, Rev. Mod. Phys. **86**, 843 (2014)
[2] P.W. Higgs, Rev. Mod. Phys. **86**, 851 (2014)
[3] H.D. Politzer, Proc. Nat. Acad. Sci. **102**, 7789 (2005)
[4] G. Aad et al., Phys. Lett. B **716**, 1 (2012)

[5] S. Chatrchyan et al., Phys. Lett. B **716**, 30 (2012)
[6] K. Riesselmann, Symmetry Magazine (2015), symmetrymagazine.org/article/july-2015/standard-model
[7] W.J. Marciano, H. Pagels, Nature **279**, 479 (1979)
[8] T. Chupp, P. Fierlinger, M. Ramsey-Musolf, J. Singh, Rev. Mod. Phys. **91**, 015001 (2019)
[9] R.E. Taylor, Rev. Mod. Phys. **63**, 573 (1991)
[10] H.W. Kendall, Rev. Mod. Phys. **63**, 597 (1991)
[11] J.I. Friedman, Rev. Mod. Phys. **63**, 615 (1991)
[12] J.I. Friedman, H.W. Kendall, R.E. Taylor, Rev. Mod. Phys. **63**, 629 (1991)
[13] T. Appelquist, R.D. Pisarski, Phys. Rev. D **23**, 2305 (1981)
[14] T.W. Appelquist, M.J. Bowick, D. Karabali, L.C.R. Wijewardhana, Phys. Rev. D **33**, 3704 (1986)
[15] A. Bashir, A. Raya, I.C. Cloet, C.D. Roberts, Phys. Rev. C **78**, 055201 (2008)
[16] A. Bashir, A. Raya, S. Sánchez-Madrigal, C.D. Roberts, Few Body Syst. **46**, 229 (2009)
[17] J. Braun, H. Gies, L. Janssen, D. Roscher, Phys. Rev. D **90**, 036002 (2014)
[18] P. Pascual, R. Tarrach, *QCD: Renormalization for the Practitioner, Lecture Notes in Physics*, Vol. 194 (Springer-Verlag, Berlin, 1984)
[19] R.L. Workman et al. (Particle Data Group), PTEP **2022**, 083C01 (2022)
[20] C.D. Roberts, S.M. Schmidt, Eur. Phys. J. ST **229**, 3319 (2020)
[21] C.D. Roberts, Symmetry **12**, 1468 (2020)
[22] C.D. Roberts, AAPPS Bulletin **31**, 6 (2021)
[23] C.D. Roberts, D.G. Richards, T. Horn, L. Chang, Prog. Part. Nucl. Phys. **120**, 103883 (2021)
[24] D. Binosi, Few Body Syst. **63**, 42 (2022)
[25] J. Papavassiliou, Chin. Phys. C **46**, 112001 (2022)
[26] M. Ding, C.D. Roberts, S.M. Schmidt (2022), *Emergence of Hadron Mass and Structure* – arXiv:2211.07763 [hep-ph]
[27] J. Carlson, A. Jaffe, A. Wiles, eds., The Millennium Prize Problems (American Mathematical Society, Providence, 2006)
[28] C.D. Roberts, Few Body Syst. **58**, 5 (2017)
[29] A.C. Aguilar et al., Eur. Phys. J. A **55**, 190 (2019)
[30] J.S. Schwinger, Phys. Rev. **125**, 397 (1962)
[31] J.S. Schwinger, Phys. Rev. **128**, 2425 (1962)
[32] J.M. Cornwall, Phys. Rev. D **26**, 1453 (1982)
[33] A.C. Aguilar, D. Binosi, J. Papavassiliou, Phys. Rev. D **78**, 025010 (2008)
[34] P. Boucaud, J. Leroy, A. Le Yaouanc, J. Micheli, O. Pene, J. Rodríguez-Quintero, JHEP **0806**, 099 (2008)
[35] D. Binosi, J. Papavassiliou, Phys. Rept. **479**, 1 (2009)
[36] P. Boucaud, J.P. Leroy, A. Le-Yaouanc, J. Micheli, O. Pene, J. Rodríguez-Quintero, Few Body Syst. **53**, 387 (2012)
[37] A.C. Aguilar, D. Binosi, J. Papavassiliou, Front. Phys. China **11**, 111203 (2016)
[38] Z.F. Cui, J.L. Zhang, D. Binosi, F. de Soto, C. Mezrag, J. Papavassiliou, C.D. Roberts, J. Rodríguez-Quintero, J. Segovia, S. Zafeiropoulos, Chin. Phys. C **44**, 083102 (2020)
[39] A. Deur, V. Burkert, J.P. Chen, W. Korsch, Particles **5**, 171 (2022)
[40] F. Wilczek, Proc. Nat. Acad. Sci. **102**, 8403 (2005)
[41] D.J. Gross, Proc. Nat. Acad. Sci. **102**, 9099 (2005)

[42] A. Pickering, *Constructing Quarks. A Sociological History of Particle Physics* (University of Chicago Press, 1999), ISBN 978-0-85224-458-6
[43] C. Itzykson, J.B. Zuber, *Quantum Field Theory* (McGraw-Hill Inc., New York, 1980)
[44] S.J. Brodsky, A.L. Deshpande, H. Gao, R.D. McKeown, C.A. Meyer, Z.E. Meziani, R.G. Milner, J. Qiu, D.G. Richards, C.D. Roberts (2015)
[45] S.J. Brodsky et al., Int. J. Mod. Phys. E **29**, 2030006 (2020)
[46] X. Chen, F.K. Guo, C.D. Roberts, R. Wang, Few Body Syst. **61**, 43 (2020)
[47] D.P. Anderle et al., Front. Phys. (Beijing) **16**, 64701 (2021)
[48] J. Arrington et al., J. Phys. G **48**, 075106 (2021)
[49] K. Aoki et al. (2021), *Extension of the J-PARC Hadron Experimental Facility: Third White Paper* – arXiv:2110.04462 [nucl-ex]
[50] C. Quintans, Few Body Syst. **63**, 72 (2022)
[51] P. Maris, C.D. Roberts, P.C. Tandy, Phys. Lett. B **420**, 267 (1998)
[52] S.J. Brodsky, M. Burkardt, I. Schmidt, Nucl. Phys. B **441**, 197 (1995)
[53] F. Yuan, Phys. Rev. D **69**, 051501 (2004)
[54] Z.F. Cui, M. Ding, J.M. Morgado, K. Raya, D. Binosi, L. Chang, J. Papavassiliou, C.D. Roberts, J. Rodríguez-Quintero, S.M. Schmidt, Eur. Phys. J. A **58**, 10 (2022)
[55] Z.F. Cui, M. Ding, J.M. Morgado, K. Raya, D. Binosi, L. Chang, F. De Soto, C.D. Roberts, J. Rodríguez-Quintero, S.M. Schmidt, Phys. Rev. D **105**, L091502 (2022)
[56] L. Chang, F. Gao, C.D. Roberts, Phys. Lett. B **829**, 137078 (2022)
[57] Y.L. Dokshitzer, Sov. Phys. JETP **46**, 641 (1977)
[58] V.N. Gribov, L.N. Lipatov, Phys. Lett. B **37**, 78 (1971)
[59] L.N. Lipatov, Sov. J. Nucl. Phys. **20**, 94 (1975)
[60] G. Altarelli, G. Parisi, Nucl. Phys. B **126**, 298 (1977)
[61] R.D. Ball, E.R. Nocera, J. Rojo, Eur. Phys. J. C **76**, 383 (2016)
[62] T.J. Hou et al., Phys. Rev. D **103**, 014013 (2021)
[63] S. Bailey, T. Cridge, L.A. Harland-Lang, A.D. Martin, R.S. Thorne, Eur. Phys. J. C **81**, 341 (2021)
[64] I. Novikov et al., Phys. Rev. D **102**, 014040 (2020)
[65] P.C. Barry, C.R. Ji, N. Sato, W. Melnitchouk, Phys. Rev. Lett. **127**, 232001 (2021)
[66] M. Aicher, A. Schäfer, W. Vogelsang, Phys. Rev. Lett. **105**, 252003 (2010)
[67] Y. Lu, L. Chang, K. Raya, C.D. Roberts, J. Rodríguez-Quintero, Phys. Lett. B **830**, 137130 (2022)
[68] E. Segarra, A. Schmidt, T. Kutz, D. Higinbotham, E. Piasetzky, M. Strikman, L. Weinstein, O. Hen, Phys. Rev. Lett. **124**, 092002 (2020)
[69] D. Abrams et al., Phys. Rev. Lett. **128**, 132003 (2022)
[70] Z.F. Cui, F. Gao, D. Binosi, L. Chang, C.D. Roberts, S.M. Schmidt, Chin. Phys. Lett. *Express* **39**, 041401 (2022)
[71] D. Carman, K. Joo, V. Mokeev, Few Body Syst. **61**, 29 (2020)
[72] V.I. Mokeev, D.S. Carman (CLAS), Rev. Mex. Fis. Suppl. **3**, 0308024 (2022)
[73] V.I. Mokeev, D.S. Carman, Few Body Syst. **63**, 59 (2022)

Chiral symmetry breaking: Current experimental status and prospects

Jan Friedrich[1,*]
on behalf of the COMPASS Collaboration
[1]Physik-Department, Technical University of Munich

Abstract. Chiral symmetry, linked to the smallness of the quark masses compared to the QCD bound states, and its breaking pattern are exploited in effective field theory to describe a multitude of phenomena by a few low-energy constants. Those concern light-meson dynamics and decays, their couplings to photons and meson-nucleon interactions. Special emphasis is given to the pion properties, in terms of pion-pion low-energy scattering, the pion polarizability and the chiral anomaly, which describes the coupling of three pions to a photon. These properties are studied by the COMPASS collaboration at CERN since first data taking with pion beams in the year 2004, and several following campaigns. In the framework of the upcoming AMBER collaboration, it is planned to extend the studies to the kaon sector.

1 Introduction

Exploring the limit of vanishing masses in quark fields of the Lagrangian of Quantum Chromodynamics (QCD),

$$\mathcal{L}_{QCD} = \overline{\psi}_q \left(i\gamma^\mu(\partial_\mu + igA_\mu) - m_q\right) \psi^q \tag{1}$$

it turns out that left- and right-handed quarks, q_L and q_R respectively, decouple and $\mathcal{L}_{QCD}$ is chirally symmetric for $m_q \to 0$. The index q denotes the different quark flavors $u, d, s, \ldots$, color indices are suppressed. This symmetry is broken explicitly by the small but non-zero quark masses, but also spontaneously in the sense that the physical vacuum does not feature the symmetry. Assuming the symmetry breaking to be a small variation, a systematic expansion of QCD at low momenta can be formulated, known as Chiral Perturbation Theory (ChPT). Pions play a central role as the (approximate) Goldstone bosons related to the chiral symmetry, and the measurement of their properties and dynamics allow to test the assumptions made in the theoretical framework, which had been pioneered by Weinberg and Goldstone, and later worked out by others [1–4]. Since ChPT is a low-momentum expansion, it converges fastest for small relative particle momenta, and at higher momenta the range of validity is constrained by higher orders within the theory, which are only calculated up to a certain order, and by the contribution of resonances, which are not included in the framework of ChPT. The domain where ChPT is relevant for the measurement depends on the concerned

*e-mail: jan@tum.de

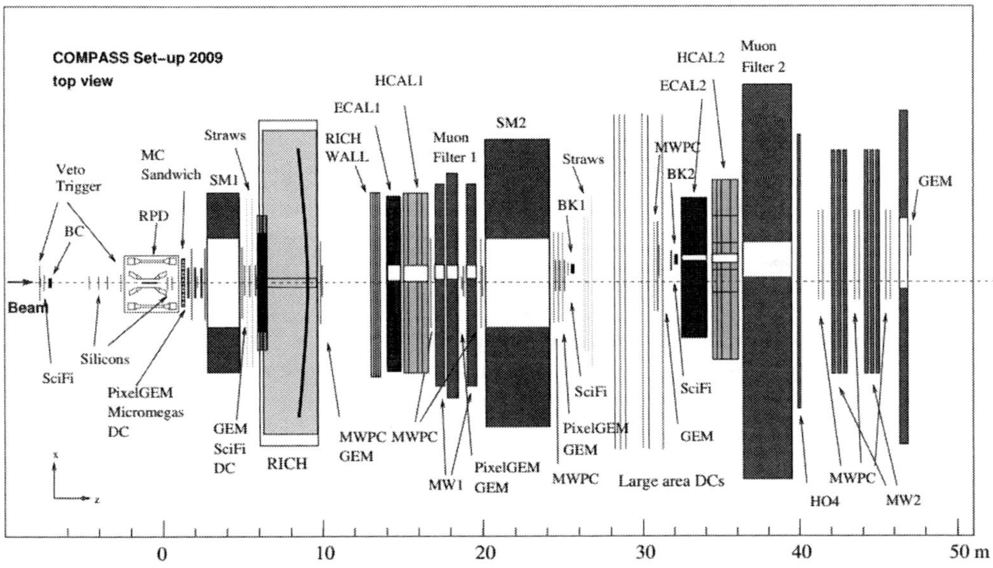

Figure 1. Layout of the COMPASS setup as used for hadron beams in the years 2008 and 2009. A detailed description of the employed detectors can be found in [7]. The spectrometer magnets SM1 and SM2, equipped up- and downstream with tracking detectors and followed by calorimeters, constitute the two stages of the setup.

process, but in general there is a range from the kinematic threshold of the process up collision energies where resonances become relevant – typically a few pion masses – where ChPT is expected to explain the dominant part of the cross sections.

The pion-pion interaction was studied by the NA48/2 Collaboration at the CERN SPS in Kaon decays, where the final-state interaction is sensitive to the pion scattering length [5], and by the E865 experiment at BNL [6].

2 Measurements at COMPASS

The COMPASS experiment uses secondary hadron and tertiary muon beams from the CERN 450 GeV super proton synchrotron (SPS). The multi-purpose detector concept allows for a wide range of investigations in hadron physics, with high-precision and high-rate capable tracking, particle identification and calorimetry in both stages of the magnetic spectrometer. The layout of the setup is shown in Fig. 1. The two stages are optimized for low and high momentum particles, respectively, and allow a momentum determination of better than 1% in a wide range, from about 1 GeV up to the beam momentum in the range of 200 GeV.

The method used to study pion-photon reactions is attributed to H. Primakoff [8]. It makes use of the intense electric field in the proximity of nuclei, which can be treated in a high-relativistic reference system as a source of quasi-real photons, to study strongly-interacting particles. The original idea concerned the measurement of the π^0 lifetime by photon-photon fusion, but it was later realized that interactions by high-energetic hadrons with the nuclear Coulomb field represent similarly a scattering off the quasi-real photon density, and consequently the whole set of such hadron interactions is referred to as Primakoff reactions.

2.1 Chiral dynamics in $\pi\gamma \to 3\pi$

An early analysis of COMPASS using the Primakoff technique was completed for the reaction $\pi^-\gamma \to \pi^-\pi^-\pi^+$ [7], of data from a pilot run in the year 2004. The region with 3-pion masses below 1 GeV is of specific interest in terms of chiral dynamics: The tails of known resonances play only a minor role, and the relative pion momenta are small, such that the kinematics lies in the region where ChPT is applicable. ChPT provides predictions for the absolute cross section to leading [9] and next-to-leading [10] order. In order to compare on the absolute level, a flux normalization for the data had to be done. For the purpose, the kaon component of the hadron beam was used, since some of the kaons decay in the free space around the target with well-known branching into the same three-pion final state as under study here, also with practically vanishing momentum transfer. Taking into account the kaon-to-pion fraction in the beam, the effective flux of pions is deduced from the number of observed kaon decays. Measuring the same final state, it features a similar reconstruction efficiency, that has to be propagated to the full mass spectrum of interest only moderately by the Monte Carlo simulation of the setup.

The result for the cross section of $\pi^-\gamma \to \pi^-\pi^-\pi^+$ is published in [11]. The absolute cross-section has been determined in five bins of the final-state mass from threshold at $3m_\pi$ up to $5m_\pi$. The data agree with the expectation from tree-level ChPT on the level of the experimental uncertainty of 20%. This confirms the extension of the ChPT approach for processes involving the coupling of four pions, *i.e.* the leading order, to processes involving the additional coupling to a photon. In terms of studying ChPT, the neutral channel $\pi^-\gamma \to \pi^-\pi^0\pi^0$ will be of even higher relevance, since for this channel higher-order loop corrections are expected to play a large role [10].

2.2 Measurement of the pion polarisability

The measurement of the pion polarisability has been one of the genuine goals in the proposal of the COMPASS experiment. After the pilot run in the year 2004, data have been collected in a two-week beam time in 2009 with significant improvements concerning the calorimetry and the trigger system, as concluded from the detailed analysis of the 2004 data. It was also found [12] that lead is not a favorable target material despite the high nuclear charge Z, since the radiative corrections due to multiple photon exchange and screening are large and represent a non-negligible source of systematic uncertainty. Consequently, the measurement was performed with a 4 mm thick nickel disk as nuclear target.

A unique feature of the pion polarisability measurement at COMPASS is that the beam can be switched, within less than an hour, from hadron to muon beam, and the spectrometer is well equipped for muon identification due to the broad physics program with muon beams. For the polarisability measurement, this allows for control measurements with muon beam, for which the theoretical expectation of the relevant bremsstrahlung process μ^- Ni$\to$ $\mu^-\gamma$ Ni is completely determined by quantum electrodynamics (QED).

The result has been published in [13], $\alpha_\pi = (2.0 \pm 0.6_{stat} \pm 0.7_{syst}) \cdot 10^{-4}$ fm^3. Radiative corrections have been applied on the level of the simulation event-wise, starting from the published calculations [14, 15] for the case of pion and muon Compton scattering, respectively, and extrapolating to the Primakoff kinematics at $Q^2 \neq 0$. Detailed information about the analysis of these data can be found in [16] The presented preliminary COMPASS value for the pion polarisability is at variance from the historically first result obtained at Serpukhov [17], also employing the Primakoff technique, which had been confirmed much later by the dedicated experiment on radiative pion photoproduction at MAMI [18]. In the mean time, the available data on $\gamma\gamma \to \pi^+\pi^-$ at e^-e^+-colliders were re-interpreted by several authors [19–22] claiming very different values for the pion polarisability. Later on, it had been proven that there is no conflict between ChPT and dispersion theory [23, 24].

So, the COMPASS result is in significant tension with the earlier experimental determinations of the pion polarisability, but it is in good agreement with the expectation of chiral perturbation theory [25].

3 Analysis of the Chiral Anomaly

The chiral anomaly constant

$$F_{3\pi} = \frac{eN_C}{12\pi^2 F_\pi^3} = (9.78 \pm 0.05)\ \text{GeV}^{-3} \quad (2)$$

describes the direct coupling of photons to three pions, as depicted in Fig. 2. In the experiment, it is accessible through the Primakoff reaction embedding $\pi^-\gamma \to \pi^-\pi^0$. The first mea-

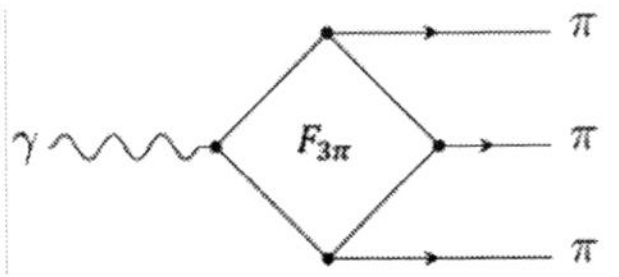

Figure 2. Coupling of a photon to three pions.

surement of $F_{3\pi}$ has been approached in the same Serpukhov experiments as mentioned above in the context of the pion polarisability. In [26], the value $F_{3\pi}^{Serp,orig} = (12.9 \pm 0.9 \pm 0.5)\ \text{GeV}^{-3}$ has been found. Later on, it was realized that a reanalysis is required due to an electromagnetic correction [27], leading to $F_{3\pi}^{Serp,corr} = (10.7 \pm 1.2)\ \text{GeV}^{-3}$. This value is in coarse agreement with the theoretical expectation, but a more precise experimental determination is highly demanded.

Theoretical framework for the analysis of the COMPASS data is the dispersive approach worked out in [28], which extends the useful kinematic range from the chiral low-momentum region beyond the ρ resonance.

The luminosity determination, which is of key importance for this analysis, has been done in principle as described above for the 3-pion data, but now extended to all relevant decay channels of the Kaons. Details of the analysis will be found in the ongoing PhD theses [29]. The preliminary value is $L_{eff} = 5.21 \pm 0.48_{syst} \pm 0.04_{stat}$, where further studies are expected to still reduce the systematic uncertainty. The level of accuracy in decribing the data by the Monte-Carlo simulation is depicted in Figs. 3 and 4 for the Kaon decays.

The most challenging part of the analysis is to control the background to the $\pi^-\gamma \to \pi^-\pi^0$ channel, which stems from reactions $\pi^-\text{Ni} \to \text{Ni}\pi^-\pi^0\pi^0$. They can proceed via Pomeron exchange and thus come with a much larger cross-section, so that even a small fraction, in which the second π^0 remains undetected and accidentally the resulting momentum transfer turns out very small, can become significant in the Primakoff region.

In order to address this background, a full analysis of $\pi^-\text{Ni} \to \text{Ni}\pi^-\pi^0\pi^0$ events is performed, and a partial-wave model fitted to the data. This model can be used as input to the Monte-Carlo simulation, extending the kinematic range into the relevant part, in which the full final state can not be detected, e.g. due to one of the π^0 having an energy too small to be reconstructed. This model can predict the amount of background differential in all kinematic quantities and thus be used to subtract the background with high accuracy. The distribution of the momentum transfer Q^2 for different 2π-mass bins is shown in Fig. 5.

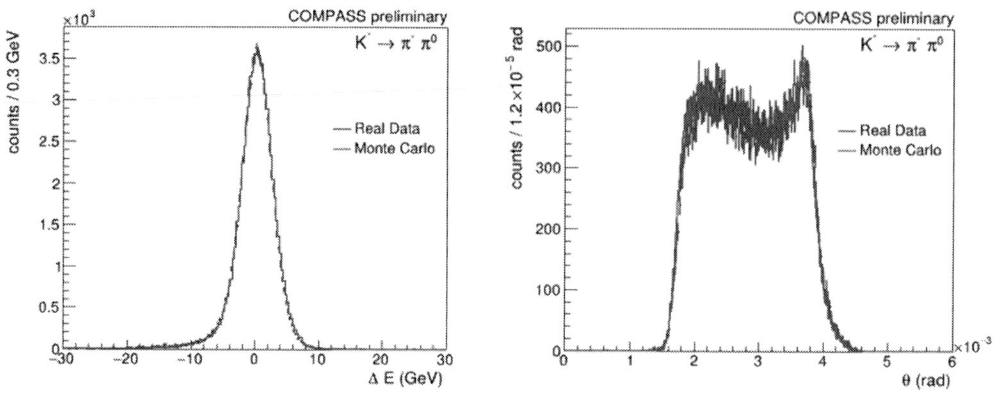

Figure 3. Comparison of the final-state laboratory energy of the Kaon decay products $\pi^-\pi^0$, minus the beam energy (left), and the angular distribution of the outgoing pions (right) with the Monte-Carlo description.

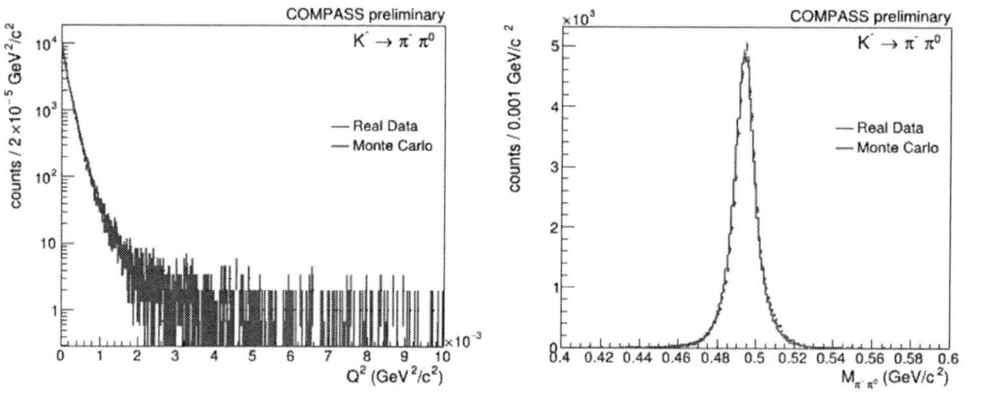

Figure 4. Comparison of the momentum balance in the Kaon decay, reconstructing a hypothetical momentum transfer, which should be zero for infinitely high detector resolution (right), and the invariant mass of the two-pion final-state system (right), with the Monte-Carlo description.

The background has a different shape than the real-data distributions, lacking the prominent Primakoff peak at small Q^2. At larger Q^2, however, it proves to be the dominant part of the distribution. In total, a consistent description of data can be achieved, thus the signal strength determined with high precision. This part of the analysis is currently still being refined.

With the preliminary background subtraction, the two-pion mass distribution shown in Fig. 6 is obtained. We conclude from fitting the model of [28] the value

$$F_{3\pi}^{COMPASS,prelim.} = (10.3 \pm 0.1_{stat} \pm 0.6_{syst}^{prelim.})\ \text{GeV}^{-3} \tag{3}$$

The result confirms the prediction of ChPT, cf. Eq. 2, with a somewhat smaller uncertainty then the earlier experiment at Serpukhov [26]. However, we aim at still significantly reducing the systematic uncertainty. Also, the QED radiative corrections, which have been worked out already, still have to be applied.

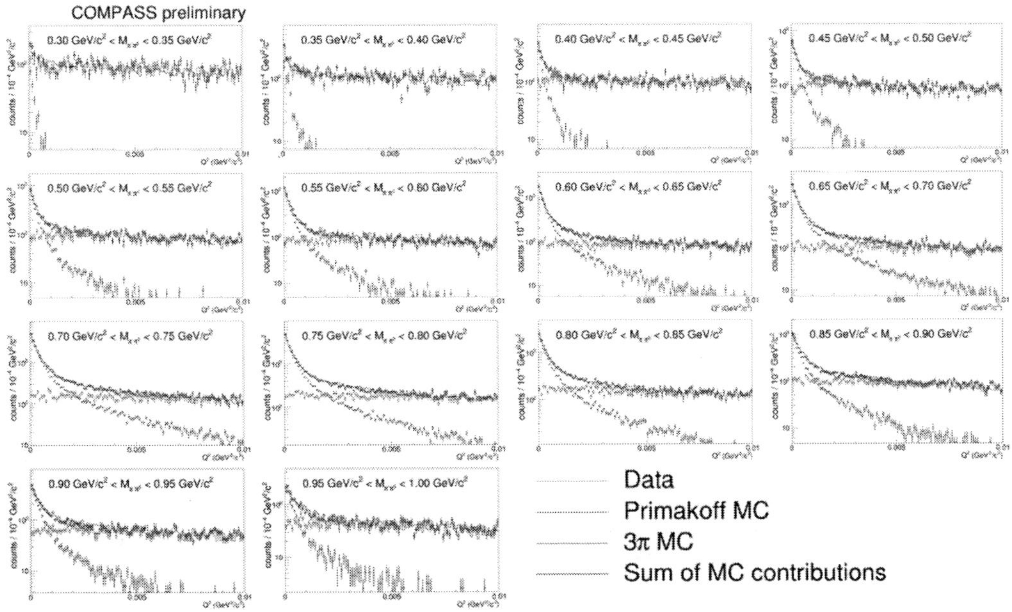

Figure 5. Three-pion background to the two-pion final states for different two-pion mass bins.

Along with analysing the data to extract $F_{3\pi}$, also the higher two-pion masses are investigated. In fact, the dispersive approach of [28] suggests to fit at the same time the contribution of the ρ resonance, and extending the two-pion mass window to above 1 GeV as shown in Fig. 6. Since the same production mechanism is at work, the reaction $\pi^-\gamma \to \rho^- \to \pi^-\pi^0$ is determined by the radiative coupling of the ρ. From the same fit that gave the result in Eq. 4, the value

$$\Gamma^{COMPASS,prelim.}_{\rho\to\pi\gamma} = (76 \pm 1_{stat}{}^{+10}_{-8,syst})\ \mathrm{keV} \tag{4}$$

is obtained. It is higher than the value (68 ± 7) keV, given in the PDG tables [30], but somewhat lower than the value (81 ± 4 ± 4) keV, obtained previously with a similar experimental method [31], however apparently neglecting the chiral contribution.

4 Conclusions and Outlook

In summary, COMPASS has contributed in several ways to test the predictions made by chiral perturbation theory as effective low-energy description of QCD. This concerns, first of all, the pion polarisability and the $\pi\gamma \to 3\pi$ transitions. Recently the analysis has been extended to the chiral anomaly $F_{3\pi}$ governing the $\pi\gamma \to 2\pi$ transition at small energies. A preliminary result has been obtained, but the investigations are ongoing especially to reduce the systematic uncertainty. The same analysis allows for a new determination of the radiative coupling of the ρ resonance.

All tests done so far with COMPASS data have confirmed the predictions of ChPT. For further challenging the theory, more precise tests are needed. COMPASS has taken more data in the year 2012, which can be analysed in the same spirit as the data from 2009, that have been shown in detail here.

On the longer range, similar studies are planned using the kaon beam component with the successor experiment of COMPASS, having been approved by the CERN SPS committee as

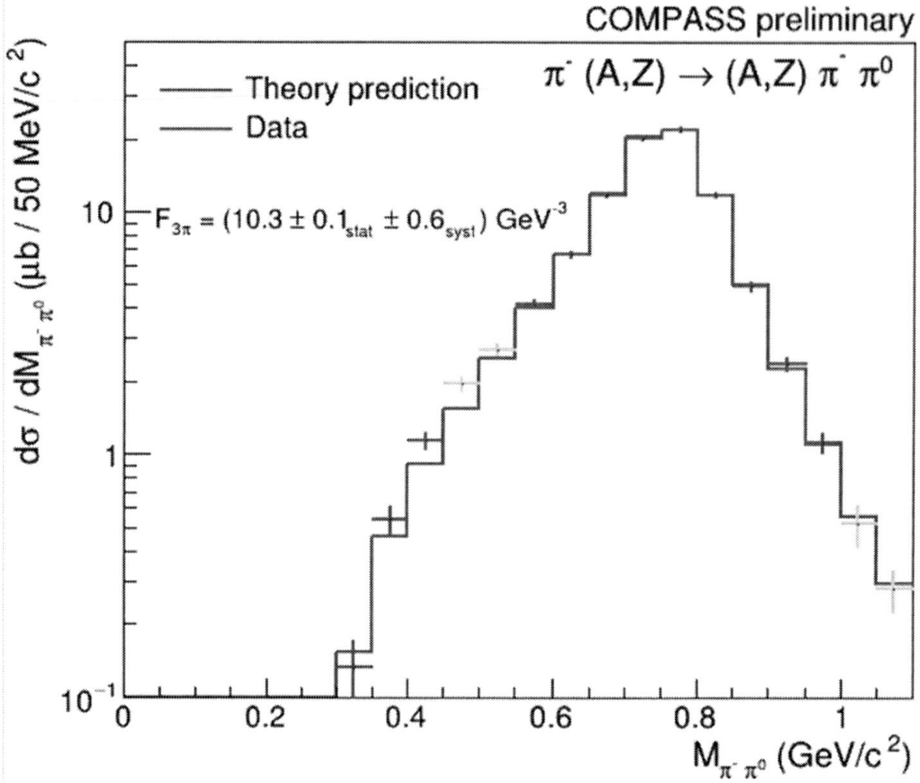

Figure 6. The preliminary COMPASS result for the measurement of the cross section of the process $\pi\gamma \to \pi\pi$, in bins of the invariant mass. The background is subtracted as explained in the text. The data are drawn in red points with their total uncertainties. The fit of the used theory framework is indicated as blue line. The region affected by decay of beam Kaons into the same final state is excluded in the fit, the respective data points shown in grey.

NA66/AMBER [32] for its Phase-1. Primakoff reactions with kaons are considered in Phase-2, and the theoretical base has recently been worked out [33] for the channel analogous to the chiral anomaly described for pions above. On the experiment side, enhancements in the beam line shall increase the number of available kaons and allow a better beam particle identification.

References

[1] J. Gasser and H. Leutwyler, Phys. Rep. **87**, 77 (1982).
[2] J. Goldstone, Nuovo Cimento **19**, 154 (1961).
[3] J. Goldstone, A. Salam and S. Weinberg, Phys. Rev. **127**, 965 (1962).
[4] S. Weinberg, Phys. Rev. Lett. **17**, 616 (1966).
[5] N. Cabibbo, Phys. Rev. Lett. **93**, 121801 (2004).
[6] S. Pislak *et al.*, Phys. Rev. D **67**, 072004 (2003), Phys. Rev. D **81**, 119903 (2010).
[7] S. Grabmüller, PhD thesis, Tech. Univ. München (2012), CERN-THESIS-2012-170.
[8] H. Primakoff, Phys. Rev. **81**, 899 (1951).
[9] N. Kaiser and J. M. Friedrich, Eur. Phys. J. A **36**, 181 (2008).

[10] N. Kaiser, Nucl. Phys. A **848**, 198 (2010).
[11] C. Adolph *et al.* [COMPASS Collaboration], Phys. Rev. Lett. **108**, 192001 (2012).
[12] N. Kaiser and J. M. Friedrich, Eur. Phys. J. A **39**, 71 (2009).
[13] C. Adolph *et al.* [COMPASS Collaboration], Phys. Rev. Lett. **114**, 062002 (2015).
[14] N. Kaiser and J. M. Friedrich, Nucl. Phys. A **812**, 186 (2008).
[15] N. Kaiser, Nucl. Phys. A **837**, 87 (2010).
[16] T. Nagel, PhD thesis, Tech. Univ. München (2012), CERN-THESIS-2012-138, http://cds.cern.ch/record/1484476.
[17] Yu. M. Antipov *et al.*, Phys. Lett. B **121**, 445 (1983).
[18] J. Ahrens *et al.*, Eur. Phys. J. A **23**, 113 (2005).
[19] D. Babusci *et al.*, Phys. Lett. B **277**, 158 (1992).
[20] J. F. Donoghue and B. R. Holstein Phys. Rev. D **48**, 137 (1993).
[21] A. E. Kaloshin and V. V. Serebryakov, Z. Phys. C **64**, 689 (1994).
[22] L. V. Fil'kov and V. L. Kashevarov, Phys. Rev. C **73**, 035210 (2006).
[23] B. Pasquini, D. Drechsel and S. Scherer, PoS of "Chiral Dynamics 09", Bern, 037 (2009).
[24] B. Pasquini, D. Drechsel and S. Scherer, Phys. Rev. C **81**, 029802 (2010).
[25] J. Gasser, M. A. Ivanov and M. E. Sainio, Nucl. Phys. B **745**, 84 (2006).
[26] Yu. M. Antipov *et al.*, Phys. Rev. D **36**, 101103 (1987).
[27] L. Ametller *et al.*, Phys. Rev. D **64**, 094009 (2001).
[28] M. Hoferichter, B. Kubis, D. Sakkas, Phys. Rev. D **86**, 116006 (2012).
[29] PhD theses of D. Ecker and A. Maltsev, in preparation.
[30] P.A. Zyla *et al.* (Particle Data Group), Prog. Theor. Exp. Phys. 2020, 083C01 (2020).
[31] L. Capraro *et al.*, Nucl. Phys. B **288**, 659 (1987).
[32] B. Adams *et al.*, AMBER Proposal, CERN-SPSC-2019-022; SPSC-P-360, http://cds.cern.ch/record/2676885.
[33] M. Dax, D. Stamen, B. Kubis, Eur. Phys. J. C **81**, 221 (2021).

The search of Axion Dark Matter with a dielectric haloscope: MADMAX

Béla Majorovits[1,*] *for the* MADMAX collaboration

[1]MPI für Physik, Föhringer Ring 6, München, Germany

Abstract. The strong CP problem is one of the most fundamental unresolved questions in the standard model of particle physics. The axion, emerging due to the maybe most elegant solution to the strong CP problem, could also solve the long standing dark matter problem. A short introduction to the strong CP problem is given. The axion as a dark matter candidate is discussed. The detection technology via inverse Primakoff effect using a dielectric haloscope is described. The MADMAX project is introduced and the status on the road towards its realization is given.

1 Introduction: The strong CP problem and the axion

The theory of Quantum Chromodynamics is based on the SU(3) group, which is non-Abelian, i.e. non commutative. This has the effect that there is an infinite amount of gauge in-equivalent degenerate zero energy states $|n\rangle$, separated by a potential barrier. Quantum mechanical tunnelling between zero energy states (Instantons) lead to the lowest energy vacuum state $|\theta\rangle$ being a superposition of all possible states $|n\rangle$ such that $|\theta\rangle = \Sigma_n e^{in\theta}|n\rangle$ with θ an arbitrary phase that the vacuum acquires. Additionally, the strong vacuum acquires a complex phase from Yukawa coupling in the weak sector. This should lead to an observable non vanishing phase of QCD: $\bar{\theta} = \theta - argdetM_q$, where M_q corresponds to the quark mass matrix. This in turns adds the CP violating term θ to the QCD Lagrangian:

$$\theta\frac{\alpha_s}{8\pi}G_{\mu\nu a}\tilde{G}_a^{\mu\nu}, \tag{1}$$

α_s being the strong coupling constant, $G_{\mu\nu a}$ and $\tilde{G}_a^{\mu\nu}$ the gluon field and its dual, respectively.

Such a non vanishing physical QCD phase should induce a measurable dipole moment of the neutron of $d_n \sim \bar{\theta} \cdot 10^{-16}\, e\,\mathrm{cm}$, where e is the unit charge. In experiments no such neutron electric dipole moment could be found [1]. The best upper limit presently is $d_n < 2 \cdot 10^{-26}\, e\,\mathrm{cm}$, which leads to an upper limit for the CP violating phase of $\bar{\theta} < 10^{-10}$. This implies that the two seemingly unrelated CP violating phases in QCD are cancelling each other to 1 in 10^{10}. This can be considered unnatural fine-tuning, as even anthropic arguments do not seem to resolve the issue. This is called the strong CP problem.

The arguably most elegant solution to this problem is to make $\bar{\theta}$ dynamical by introduction of a chiral U(1) Peccei Quinn (PQ) symmetry, spontaneously broken by the vacuum expectation value of a complex scalar field at an energy scale f_a, also called the axion decay

*e-mail: bela@mpp.mpg.de

constant [2]. Here, $\bar{\theta}$ is the phase of the scalar field. After spontaneous symmetry breaking the vacuum obtains an arbitrary phase θ_i. During and after the QCD phase transition the PQ symmetry is also explicitly broken by the topological susceptibility of the QCD vacuum. Like this, the phase $\bar{\theta}$ obtains a non vanishing potential, with its minimum around zero, which is CP conserving. Hence, after the QCD phase transition $\bar{\theta}$ can start oscillating around the minimum. The higher θ_i, the more energy will be stored in the oscillation of the axion field. The oscillation is damped by the expansion rate of the universe during the phase transition, hence being dynamically driven to a value very close but different from zero. The oscillation of $\bar{\theta}$ is the axion [3]. The axion mass m_a mass is inversely proportional to f_a and can be calculated in QCD as

$$m_a \sim 5.7\mu eV \frac{10^{12} GeV}{f_a}. \tag{2}$$

It turns out that the axion is also an ideal cold dark matter candidate. It is produced non thermally, hence should be non relativistic in the present universe. If it is assumed that axions make up the whole dark matter density of the universe $\rho_{DM} = 1.2\,\mathrm{keV/cm^3}$, some predictions can be made for the axion mass. For the case that PQ symmetry breaking happened before the inflationary epoch of the universe, today's observable universe was in casual contact during the PQ phase transition and relation can be given between $|\theta_i|$ and m_a. As $|\theta_i|$ is arbitrary, m_a in this case is predicted to range from ~ 1 neV for $|\theta_i| = 0.01$ to ~ 1 meV for $|\theta_i| = \pi$. If PQ symmetry breaking happened after the inflationary epoch, today's observable universe would consist of many patches non-causually connected during PQ symmetry breaking. Hence, the observed dark matter density ρ_{DM} needs to be related to the average $|\theta_i|$ from all patches causally disconnected during the phase transition. This, in principle could give an accurate prediction for m_a. However, in this case topological defects like strings and domain walls contain extra energy that later decays into an additional population of axions. The calculation of the additional amount of axions due to the topological defects brings in considerable uncertainties. The prediction for m_a in the post-inflationary scenarios is thus widened to a range between $\sim 25\,\mu$eV and ~ 100 meV.

Axions can couple to photons via mixing with neutral mesons, like the π^0. Hence, it can in principle decay to two photons. The coupling is, however, suppressed by f_a, leading to a half life many orders of magnitudes longer than the age of the universe for the relevant axion mass. The "inverse Primakoff effect" can be used to convert axions to photons: a B-field "supplies" a virtual photon to the axion, which can then be converted into a single photon. Hence, axion field oscillations inside a B-field induce tiny E-field fluctuations. Again, the conversion probability is suppressed by f_a. The amplitude of the induced E-field oscillation will be proportional to the B-field, hence the power of the radiation that converts from axion oscillations to E-field oscillations scales with B^2.

Any spontaneously and explicitly broken U(1) symmetry is actually leading to a massive Goldstone Boson. In fact, string theory predicts a plethora of such particles from compactification. This is sometimes also called the 'axiverse'. Here, no relation between mass of the axion like particle (ALP) and energy scale of spontaneous symmetry breaking can be given and prediction of the viable mass range is even more relaxed as for axions.

It should be noted here, that some astrophysical anomalies such as the anomalous cooling of stars [4] or the anomalous transparency of the intergalactic medium [5] could actually be explained by ALPs with $g_{a\gamma} \sim 10^{-10}\,\mathrm{GeV^{-1}}$.

2 Detection of axions and ALPs: cavity and dielectric haloscope

Axion or ALP induced E-field oscillations in a static magnetic field can be amplified if they happen inside a resonant cavity that is tuned to the frequency of the oscillations. This is uti-

https://doi.org/10.1051/epjconf/202328201008

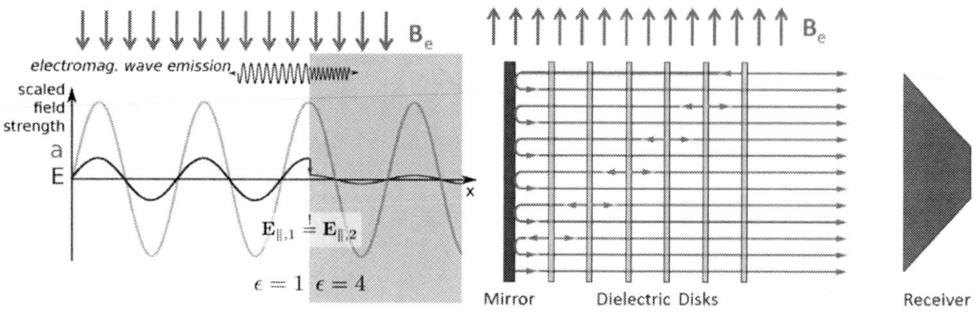

Figure 1. Left: Visualization of photon emission from surfaces with change of dielectric constant. Right: Sketch of the principle of a dielectric haloscope. For details see text.x

lized by the experiments presently leading the field, ADMX [6] and the CAPP [7] initiatives. These experiments use cavities with high quality factor. They utilize a low loss dielectric tuning rod inside the cavity, that changes the resonant frequency of the cavity when moved. This allows for scanning a wide mass range. These experiments have managed to reach a sensitivity covering a large fraction of the benchmark models compatible with theory predictions [8, 9]. They are sensitive in the mass range between ~ 1 and $\sim 40\,\mu$eV and have set limits in the $2\,\mu$eV mass range.

Their sensitivity scales with the volume of the cavity, hence the range in mass is limited by the size of the cavity used. For resonant conditions the diameter of the cavity roughly has to match the half wavelength of the axion induced E-field oscillations. For higher frequencies, hence smaller cavities need to be used. Additionally, the Q-factor of cavities is deteriorating for higher frequencies due to the skin effect. This so far limits the axion mass range accessible by the cavity haloscope experiments.

The dielectric haloscope approach has been introduced to circumvent the high frequency limitations of the cavity approach [10]. It utilizes the effect that at boundaries between two media with different dielectric constant with a static B-field parallel to the surface, the axion induced E-field oscillation has a discontinuity, as shown in Fig. 1(left). To compensate for this, electromagnetic waves are emitted from the surface [11]. Assuming a local axion dark matter density of $\rho_{DM} = 0.3\,\mathrm{GeV/cm^3}$, for a perfect mirror ($\epsilon = \infty$) this leads to an emission of EM radiation with a power of

$$P \sim 2 \cdot 10^{-27} \left(\frac{B}{10\mathrm{T}}\right)^2 \left(\frac{A}{1\mathrm{m}^2}\right) C_{a\gamma}^2 \;\mathrm{W}\,, \tag{3}$$

with $C_{a\gamma} \sim O(1)$ for axion benchmark models [8, 9].

In order to boost the emitted power, the dish antenna approach can be extended by putting discs from low loss dielectrics in front of a perfect mirror. Like this, there will be coherent emission from all surfaces of the discs and the mirror, additionally the coherent radiation from all surfaces can constructively (or destructively) interfere, as displayed in Fig. 1(right). With proper spacing between the individual discs and the mirror, a boost of the emitted power in a quasi broad frequency range can be obtained [12]. The frequency and the bandwidth of the boost factor depend on the disc and mirror configuration. With adjustable disc distances, hence the axion mass can be scanned. This concept can be used to access the so far unexplored axion mass range around $100\,\mu$eV. The discs and mirror configuration leading to the actual power boost of the signal is called "Booster" in the following.

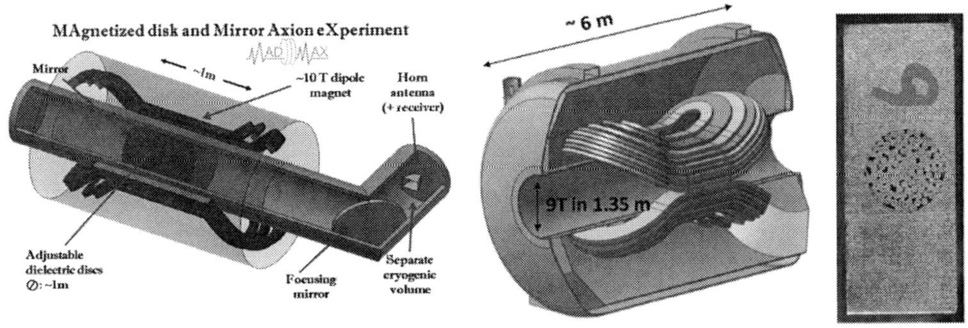

Figure 2. Left: Sketch of MADMAX baseline design. Center: Drawing of the planned MADMAX magnet configuration. Right: CICC supeconductor cross section after compaction.

3 MADMAX status of R&D

MADMAX is based on the dielectric haloscope concept. As baseline design it is envisioned to place a booster with many dielectric discs of area $\sim 1\ \mathrm{m}^2$ inside a large warm bore dipole magnet with B-field $O(10\mathrm{T})$. A signal power of a few $10^{-23}\ W$ is needed as this can be detected by state of the art detectors. For such a configuration a power boost factor of roughly 10^4 is needed. This could be obtained for example by a set of some tens of discs [13]. A sketch of the baseline design is shown in Fig. 2.

Hall north of the former HERA ring at DESY is the designated site of MADMAX. This comes with the advantage of being able to use the DESY infrastructure, for example for cryogenics. Also, MADMAX will reuse the yoke of the former H1 detector solenoid [14]. Its use ensures that the fringe field in the direct vicinity of the experiment is low enough and in parallel homogenises and even increases the B-field in the centre of the magnet.

The required magnet is presently being developed and designed by CEA Irfu in cooperation with Bilfinger NOELL in the framework of an innovation partnership (EU tendering tool). The design is based on 2 times 9 skateboard double pan cake coils. A sketch of the magnet design is shown in Fig. 2 (centre). The coils are made from a novel conductor in conduit cable (CICC) design, where the superconducting filament made from NbTi is guided inside a hole of a copper conduit. Cooling of the superconductor happens via superfluid helium at 1.8 K inside the hole of the conductor. By now it has been verified that the required components to build such a conductor can be purchased from industrial partners. Also, it has been demonstrated that the conductor can be produced in the required length with the needed compaction leading to the proper ratio of superconductor to void area inside the hole of the stabilizer. A cross section of the conductor after compaction is shown in Fig. 2 (right).

Another important question that needed to be answered was the possibility of quench detection, i.e. the propagation velocity of quenches inside the detector. For safe magnet operation, a possible quench needs to be detected within 0.1 s, which translates into a minimum quench propagation velocity of 1 m/s. To test this, a dedicated solenoid using a CICC cable according to MADMAX specifications has been designed and built by the two innovation partners. It has been extensively tested inside a test vacuum cryostat at CEA. The solenoid was equipped with proper sensors on top of the conductor, allowing to measure the velocity of quenches. Controlled quenches could be triggered using a heater. Like this it could be demonstrated that the MADMAX CICC superconductor has a quench velocity compatible with safe quench protection. Figure 3 (left) shows the measured quench velocity and the beginning of the quench as a function of the applied current to the coil. Clearly, even for an applied

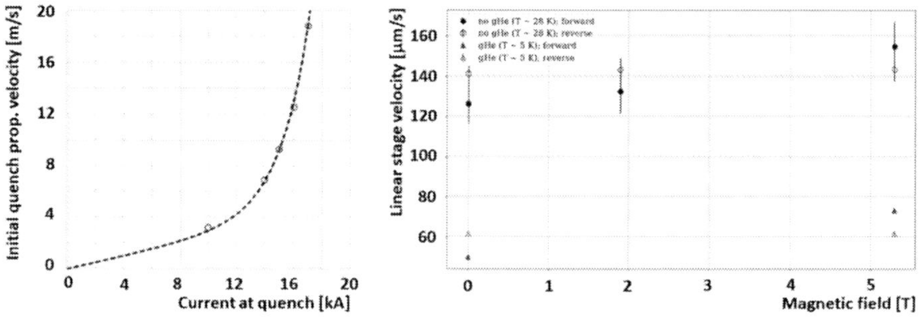

Figure 3. Left: Quench propagation velocity as function of applied voltage. Right: Speed of piezo motor carriage under different B-fields and helium gas pressure.

current as low as 10 kA (operational current at 21 kA), the minimum quench propagation velocity of 1 m/s is exceeded.

One of the most important questions is how to determine and calibrate the actual power boost factor. Reflection measurements of the setup give important information on the optical behaviour of a booster itself. However, they do not give direct information on the signal shape expected from coherent emission of all disc and mirror surfaces. Also, the influence of the first stage amplifier itself is not obtained. To overcome these shortcomings, additional noise measurements of the system are performed. These provide the interference between the amplifier and the booster system as well as the thermal noise from the booster system itself, which is emitted directly from the booster disc and mirror surfaces, similar as expected from axion induced radiation. A 1D noise model with input parameters taken from the reflectivity measurements can then be fitted to the actual noise measurement itself. Like this, information on the relevant parameters for determination of the power boost factor can be extracted. By varying the parameters within their uncertainties, the uncertainty of the power boost factor determination can be obtained.

To verify this method that allows for a reliable boost factor determination, a reduced booster with fixed disc and mirror positions with well defined boundary conditions was designed and built. Three sapphire discs with 100 mm diameter were put in front of a copper mirror inside a tight metallic cylinder. A schematic of the setup in explosion is shown in Fig. 4 (left). The simulated RF response of the system in the frequency regime between 18 and 20 GHz did well reproduce the measurements. Also the established noise model could well describe the actual noise measurements. The shape of the power boost factor with maxiumum at 750 obtained from a corresponding 1D model is shown in Fig. 5 (right) as overlay to a background measurement.

A further challenge of the MADMAX design concept is the actuation over macroscopic distances (up to a few cm) of the many discs with the necessary precision (around μm) in the cryogenic environment at high B-field. From the beginning piezo motors had been identified as the most promising technology. An R&D project had been started together with JPE [1]. The design foresees three motor carriages per disc which are actuated by a piezo crystal via stick

[1] https://www.jpe-innovations.com/

 https://doi.org/10.1051/epjconf/202328201008

& slip. Controlling of the positions and tilts of the discs is done using laser interferometry, measuring the distances for each disc between the referenced copper mirror and the three points on the individual disc rings.

As a first step, a motor carriage equipped with a piezo crystal was designed and produced. It is running on a high precision cryo compatible ceramic rail. The system was tested at cryogenic temperature (~ 5 K) in vacuum, as well as in a test magnet with a B-field of 5.3 T applied. Latter tests were also performed at a temperature of 5 K in an exchange gas helium surrounding (10^{-2} mbar). The motor was reliably operated and the obtainable speed was measured to be around $100\,\mu$m/s. Figure 3 shows the measured velocities in the forward and backward directions for different settings (no gHe: vacuum, $< 10^{-4}$ mbar, gHe: helium exchange gas at $\sim 10^{-2}$ mbar) and with different B-field applied. These tests show that piezo motors are suitable for use in the MADMAX setup.

To show the feasibility of building the MADMAX booster with the needed precision, a mechanical mock up with a mirror and one disc was built. The diameter of mirror and disc were 200 mm. The mock up - called project 200 - does include all mechanical parts that are necessary to build the MADMAX booster according to its baseline design. Most importantly, the interplay between piezo motors, the disc and the interferometer system can be tested in different surroundings, i.e. at cryogenic temperatures and inside a B-field. After precision assembly and commissioning at room temperature the setup has been tested inside a vacuum cryostat at CERN. In both surroundings, the discs could be reliably moved and the disc positions controlled to the needed precision of better $\sim 1\,\mu$m. Also, the setup was reliably and reproducibly operated inside the 1.6 T B-field of the MORPURGO magnet at CERN [15].

A Receiver DAQ system has been developed earlier to measure the power output of the CB100 system in the relevant frequency regime. It is based on a low noise first stage amplifier coupled to a heterodyne mixing scheme with three intermediate frequencies and four samplers with on board FPGAs that are recording the fast Fourier Transforms with a dead time of $< 2\%$. The setup with CB100 attached has been tested inside the 1.6 T B-field of the MORPURGO magnet at CERN in the north area. It could be shown that the highly sensitive receiver system could be operated in the direct surrounding of the magnet. Radio frequency interference measurements have revealed that no spurious lines appeared and that the system has a sensitivity as expected from earlier investigations. Figure 5 (left) shows the whole

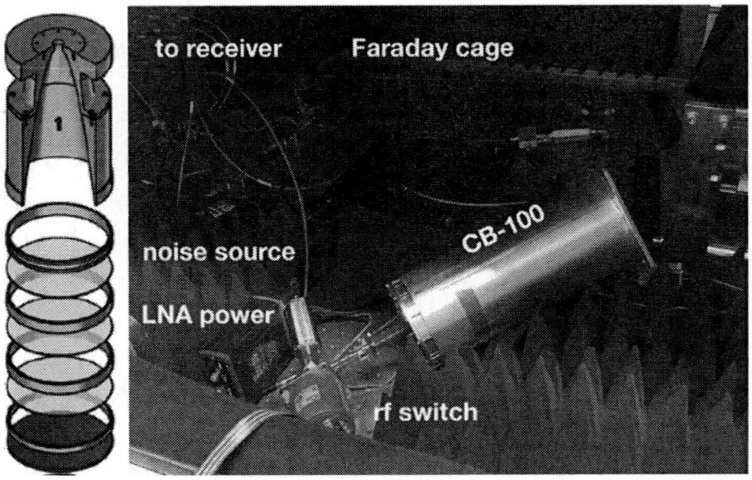

Figure 4. Left: Explosion drawing of CB100. Right: CB100 inside Faraday cage.

setup at the MORPURGO magnet. Figure 5 (right) displays the 10 hour data set taken with CB100 inside the MORPURGO magnet with the B-field at 1.6 T. Overlayed is the boost factor of CB100, indicating at which frequency the sensitivity is the highest due to the potential signal receiving a power boost proportional to the black curve. The center frequency is at roughly 19.65 GHz, corresponding to $\sim 78.6\,\mu$eV. With the obtained sensitivity at the center frequency, ALP photon couplings $g_{a\gamma} > 10^{-10}\,\mathrm{GeV}^{-1}$ can be excluded.

4 Conclusion and Outlook

Axions have been proposed to explain the strong CP problem. They are also an excellent candidate to solve the dark matter problem. The dielectric haloscope concept has the potential to detect dark matter axions in the mass range compatible with the post inflationary PQ symmetry breaking scenario around $100\,\mu$eV. The MADMAX project is developing the dielectric haloscope concept with the goal to achieve the required sensitivity. Significant milestones have lately been reached. A method to determine the power boost factor of the axion induced E-field including its uncertainty has been shown to work on a closed booster system. Piezo motors have been verified to work in strong B-field and at cryogenic temperatures. The conductor for the required magnet has been shown to be suitable for safe operation in terms of fast enough quench protection. Industrial suppliers have been identified that can deliver the required conductor and conduit with required tolerances. First measurements of a booster setup have been performed at CERN inside the MORPURGO magnet. The measurements show that radio frequency interference is not deteriorating the sensitivity of the setup.

The projection of the sensitivity of MADMAX as a dielectric haloscope is shown in Fig. 6 together with existing limits from cavity experiments as well as from the CAST solar axion experiment. Note that the haloscope limits and projections are based on the assumption that axions explain all of the dark matter density. Also shown in the plot are projections for future solar experiments and the ALPS II light shining through the wall experiment. Furthermore, the regions compatible with the astrophysical hints are indicated. QCD axions are predicted to lie on the diagonal line according to the most familiar KSVZ and DFSZ benchmark models. MADMAX could cover a significant fraction of the parameter space predicted for the model in which Peccei Quinn symmetry is broken after cosmic inflation.

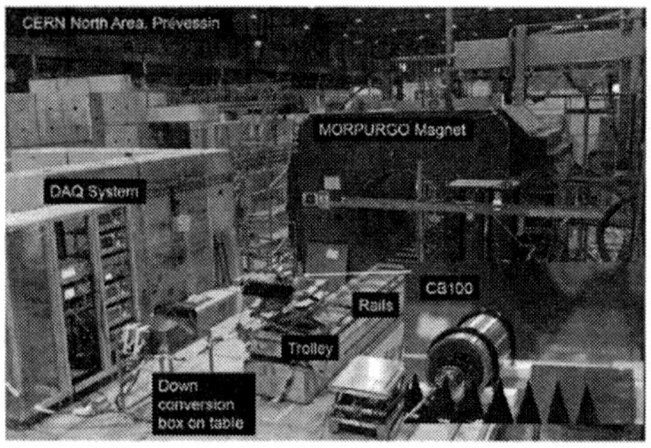

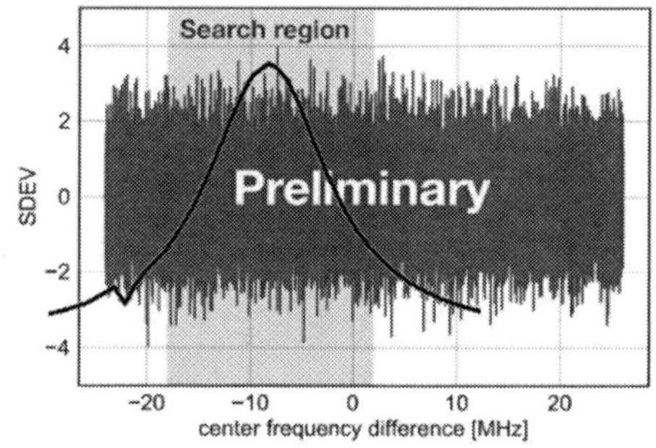

Figure 5. Left: CB100 setup including receiver and DAQ at CERN in front of the MORPURGO magnet before insertion. Right: Background measurement with the receiver and DAQ system with CB100 connected and B-field on. The measured power is normalized to the noise level in terms of standard deviation. The black curve gives the shape of the power boost factor of CB100. The shaded area indicates the frequency range where the setup is sensitive due to the power boost.

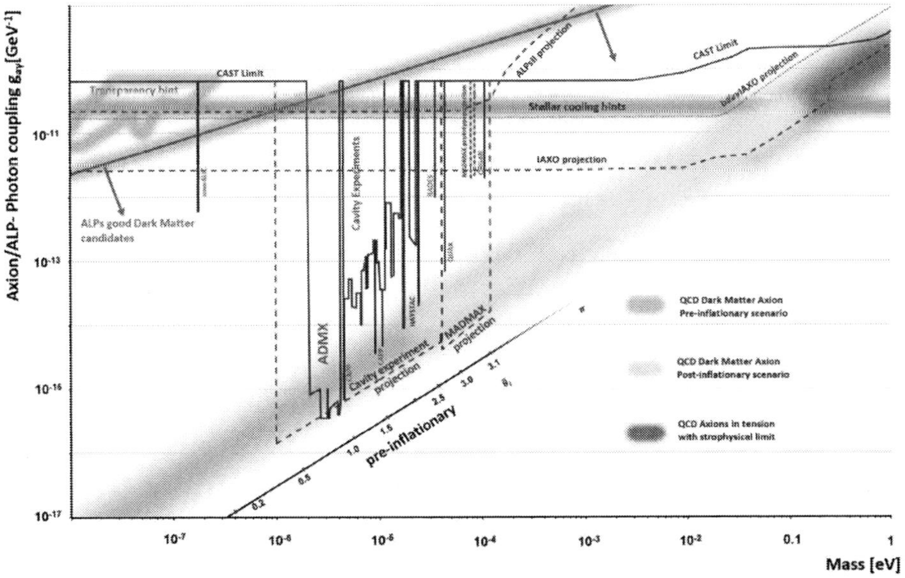

Figure 6. Plot of the axion/ALP to photon coupling vs. mass. Limits are indicated as solid lines, projections as dashed lines. Astrophysical limits are also shown as shaded areas in orange. The diagonal gives the range compatible with KSVZ and DFSZ axions.

References

[1] C. Abel et al., Phys. Rev. Lett. **124** 081803 (2020)

[2] R.D. Peccei and H. Quinn, Phys. Rev. Lett. **38** 1440 (1977) and Phys. Rev. D **16** 1791 (1977)

[3] S. Weinberg, Phys. Rev. Lett. **40** 223 (1978) and F. Wilczek, Phys. Rev. Lett., **40** 279 (1978)

[4] M. Giannotti et al., JCAP **1605** 057 (2016) and M. Giannotti et al., JCAP **1710** 010 (2017)

[5] D. Horns and M. Meyer, JCAP **1202** 033 (2012) and P. Brun, J. Phys. Conf. Series **460** 012015 (2013)

[6] ADMX collaboration, C. Bartram et al., Phys. Rev. Lett. **127** 261803 (2021)

[7] A.K. Yi et al., arxiv:[2210.10961] and J. Kim et al. arxiv:[2207.13597]

[8] J.E. Kim, Phys. Rev. Lett. **43** 103 (1979) and M.A. Shifman, A.I. Vainshtein and V.I. Zakharov, Nucl. Phys. B **166** 493 (1980)

[9] M. Dine, W. Fischler and M. Srednicki, Phys. Lett. B **104** 199 (1981) and A.R. Zhitnitsky, Sov. J. Nucl. Phys. **31** 260 (1980)

[10] MADMAX interest group, A. Caldwell et al., Phys. Rev. Lett. **118** 091801 (2017)

[11] D. Horns et al., JCAP **1304** 016 (2013)

[12] J. Jaeckel and J. Redondo, Phys. Rev. D **88** 11502 (2013)

[13] MADMAX collaboration, P. Brun et al., Eur. Phys. J. C **79** 186 (2019)

[14] H1 collaboration, I. Abt et al., Nucl. Inst. Meth. A **386** 310 (1997)

[15] M. Morpurgo, Cryogenics **19** 411 (1979)

Observation of the $\gamma\gamma \rightarrow \tau\tau$ process in Pb+Pb collisions and constraints on the τ-lepton anomalous magnetic moment with the ATLAS detector at LHC

Vincenzo Cavasinni[1]
on behalf of the ATLAS Collaboration

[1]Dipartimento di Fisica "E.Fermi" dell'Università and INFN Pisa

Abstract. This paper reports the observation of τ-lepton pair production in ultraperipheral lead-lead collisions, $Pb(\gamma)Pb(\gamma) \rightarrow \tau\tau$ and constraints on the τ-lepton anomalous magnetic moment, a_τ, measured by the ATLAS experiment. The dataset corresponds to an integrated luminosity of 1.44 nb^{-1} of LHC Pb+Pb collisions at $\sqrt{s_{NN}}$=5.02 TeV. Selected events contain one muon from a τ-lepton decay, an electron or charged-particle track(s) from the other τ-lepton decay, little additional central-detector activity, and no forward neutrons. The $\gamma\gamma \rightarrow \tau\tau$ process is observed in Pb+Pb collisions with a signal strength of $\mu_{\tau\tau}$= 1.03 $^{+0.06}_{-0.05}$. To measure a_τ, a template fit to the muon transverse-momentum distribution from τ-lepton candidates is performed, using a dimuon ($\gamma\gamma \rightarrow \mu\mu$) control sample to constrain systematic uncertainties. The observed 95% confidence-level interval for a_τ is $(-0.057, 0.024)$.

1 Introduction

The magnetic moment μ_B is proportional to the particle spin through the parameter g: $\mu_B = g\frac{e\hbar}{2m}$. In the Dirac theory $g = 2$ and the anomalous magnetic moment a is defined as $a = \frac{g-2}{2}$. The first QED correction to Dirac theory for electrons was calculated by J. Schwinger in 1948: $a_e = \alpha/2\pi = 0.00116$ [1]. In the Standard Model (SM) higher order corrections predict a to be of the order of 10^{-3} calculated with high precision that requires to be verified with corresponding accurate measurements. Discrepancies between the SM predictions and the measurements would suggest new physics beyond SM, eg: lepton- compositeness, supersymmetric particle, ...

For electrons and muons the measurements of a have been performed with a precision of 0.28 ppb [2] and 0.46 ppm [3] with differences from SM predictions of 2.4 and 4.6 standard deviations, respectively. Therefore the precise measurements of the lepton anomalous moment represent one of the most powerful tools to check the validity of the SM in searching for new physics.

For a_τ, at the present, the experimental measurement is much less precise than those possible for the other two charged leptons. On the other hand, theoretically, the larger τ−mass enhances the contributions to a from new particles of mass M; in fact the photon-lepton vertex coupling is expected to be of the order of m_l^2/M^2 for a lepton of mass m_l. Therefore new physics effects for τ's would be larger by a factor $m_\tau^2/m_\mu^2 = 286$ compared with the case of muons. The present experimental knowledge of a_τ is quite poor: the very short τ-lifetime

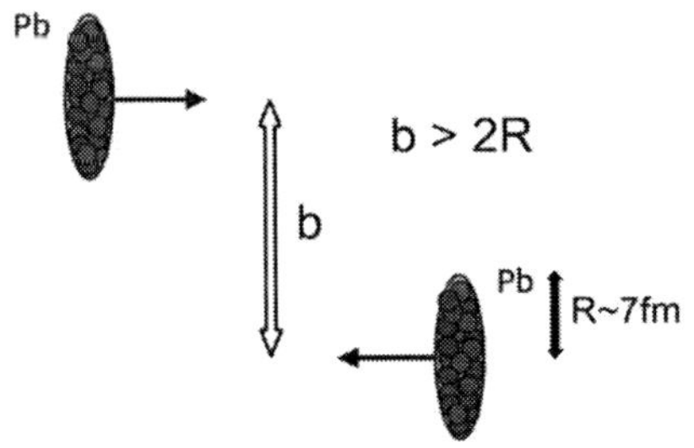

Figure 1. Example of Pb-Pb ultraperipheral collision. Pb-Pb collisions occurring at an impact parameter b large, say twice the Pb nucleus radius, are not affected by strong nucleon-nucleon interactions and they are called Ultra Peripheral Collisions.

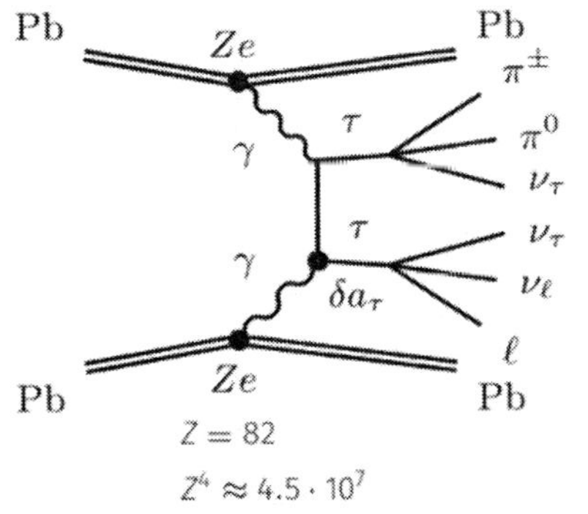

Figure 2. Diagram of pair-production of tau leptons from ultraperipheral Pb-ion collisions. One τ decays into μ + neutrinos, the other τ decays into 3 pions + neutrino. New physics modifies the magnetic moment by δa_τ.

precludes the use of the precession frequency measurement method as done in the μ−case. The method adopted was then to measure the $\tau - pair$ production cross-section, total and differential, in photon-photon scattering, where these quantities are sensitive to a_τ. The best a_τ measurement was obtained by the DELPHI experiment at LEP [4]:$-0.052 < a_\tau^{exp} < 0.013$ at 95% CL, or, expressed as a central value and an error, $a_\tau^{exp} = -0.018(17)$.

The theoretical SM expectation for a_τ , which includes electroweak and hadronic contributions, is calculated in [5]: $a_\tau^{SM} = 117721(5) \times 10^{-8}$ with a precision by far better than that of the experiments. So any effort should be made experimentally to approach in precision at least the size of the prediction of the SM: $a_\tau = 10^{-3}$.

The ATLAS experiment [6] has measured a_τ using Pb-Pb ultra peripheral collisions (UPC) at LHC to single-out $\gamma - \gamma$ collisions yielding tau-pairs. This reaction has several advantages compared with other reactions at LHC. In fact in UPC Pb-Pb collisions, Figure 1, the τ-pairs are generated by a $\gamma\gamma$ collision [1], Figure 2. The cross-section is strongly enhanced by a factor Z^4, largely compensating the lower luminosity compared with that available in proton-proton collisions. In addition, the request of the presence of only the tau-decay products in the detector, with essentially no pile-up background, makes the control of backgrounds (electron/muon/quark pairs) much easier than in case of p-p collisions.

2 The analysis strategy

This paper presents the observation of the $Pb(\gamma)Pb(\gamma)) \rightarrow \tau\tau$ process and measurement of a_τ [6] using 1.44 nb^{-1} integrated luminosity at $\sqrt{s_{NN}} = 5.02$ TeV Pb+Pb data recorded by ATLAS in 2018.

The a_τ measurement in ATLAS is performed by looking at the dependence of the $Pb(\gamma)Pb(\gamma) \rightarrow \tau\tau$ cross-section and τ-p_T distribution. Figure 3 [7] shows the a_τ dependences of cross-section, left plot, and of τ-p_T distribution, central and right plots, for negative and positive a_τ, respectively.

[1]The first idea of using photons accompanying fast moving charged particles for physics experiments is due to Enrico Fermi in 1924 [8].

The ATLAS detector [9] is a multipurpose particle detector with cylindrical geometry, comprising an inner detector (ID) tracker, electromagnetic (EM), hadronic calorimeters and a muon spectrometer (MS). The zero-degree calorimeters (ZDC) [10] are located at z = 140 m in both sides from the interaction point and detect neutral particles such as neutrons emitted from interacting nuclei.

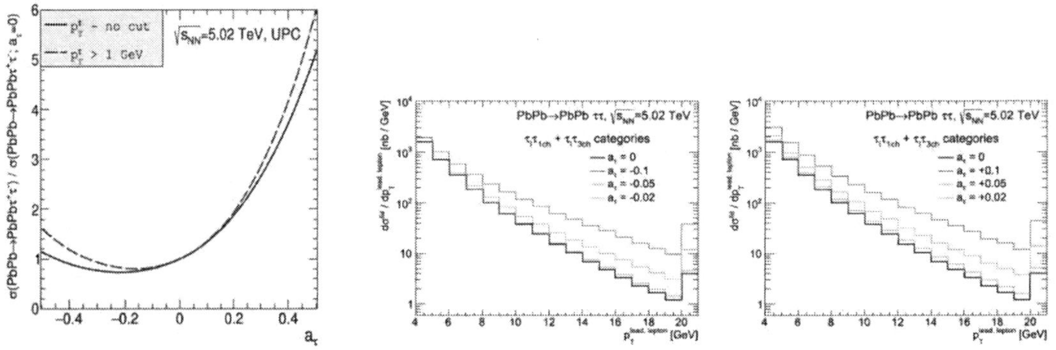

Figure 3. τ-pair cross-section dependence on a_τ (normalised to a_τ = 0) (left), and τ-p_T distribution as a function of positive (central) and negative (right) a_τ[7].

After the UPC Pb-Pb interaction, the ions can remain intact, enabling selection of low-multiplicity events with one muon originating from one of the τ-leptons, while the other τ-lepton decay is reconstructed as either an electron or one or three charged-particle tracks with low transverse momentum. The trigger required one muon with $p_T > 4$ GeV, low activity in the forward calorimeters and no neutron in the very forward calorimeter ZDC.

Samples of simulated $\gamma\gamma \to \tau\tau$ signal events were produced at leading order in QED using the Starlight 2.0 [11] Monte Carlo (MC) generator for lepton-pair production in UPC events, interfaced with Tauola [12] for τ-lepton decays. The dominant background sources is the $\gamma\gamma \to \mu\mu$ process, its contribution is estimated with the aid of MC samples generated using Starlight; Pythia 8 was used to model EM FSR from the muons.

Selected events must contain one muon, which targets a muonic decay of one of the τ leptons. Three signal regions (SR) then categorise events by the decay signature of the other τ-lepton. The μe-SR category additionally requires one electron, and no additional tracks which targets fully leptonic decays of both τ-leptons. The $\mu 1T$-SR ($\mu 3T$-SR) category requires exactly one track (three tracks) separated from the muon by $\Delta R(\mu - trk) > 0.1$ [2], which targets τ-lepton decays to one or three charged hadrons. The one-track requirement also captures leptonic τ-lepton decays that fail electron or muon reconstruction. The electric charges of muon, electron, and tracks must sum to zero. Figure 4 shows the display of a candidate signal event in $\mu 3T$-SR.

To constrain the $\gamma\gamma \to \mu\mu$ a control region (CR) of dimuon events called 2μ-CR is defined. It requires exactly two muons with invariant mass above 11 GeV to suppress quarkonia ($Y_{nS} \to \mu\mu$) backgrounds and no additional tracks separated from the muons by $\Delta R(\mu - trk) > 0.1$. The 2μ-CR sample was used also to constrain the systematic uncertainty arising from the knowledge of $\gamma\gamma$ flux.

[2] $\Delta R(\mu - trk) = \sqrt{(\eta_\mu - \eta_{trk})^2 + (\Phi_\mu - \Phi_{trk})^2}$ where η_μ, Φ_μ and η_{trk}, Φ_{trk} are the pseudorapidities and the azimuthal angles of the muon and of the track, respectively.

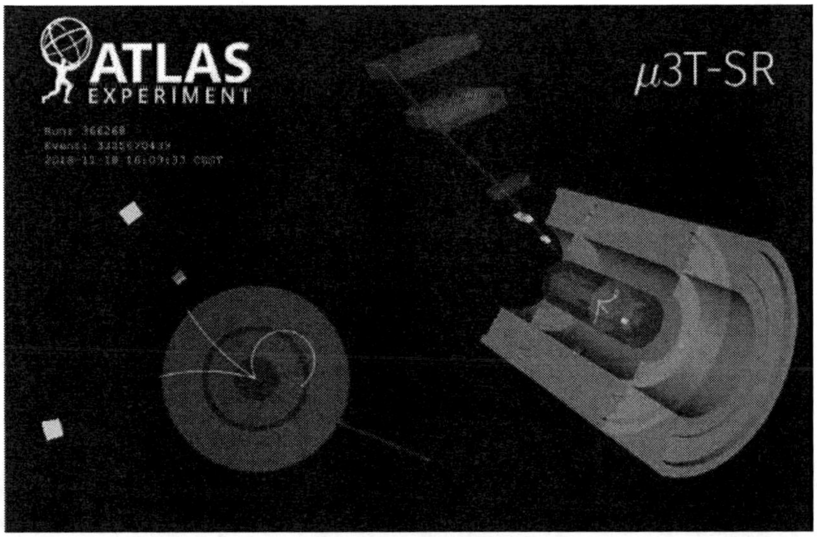

Figure 4. Display of a candidate event in the central ATLAS detector: a muon (red track) + 3 tracks. The muon is measured by the muon-spectrometer while the tracks are measured by the central ATLAS tracker [6].

The second most important background, significant only for μ1T-SR and μ3T-SR, comes from the diffractive photo-nuclear events where the Pb-ions may not dissociate failing the ZDC neutron veto. This background was estimated by a data-driven method using dedicated CRs called μ2T-CR and μ 4T-CR, which apply the same selection as μ1T-SR and μ3T-SR, but require an additional low energy track satisfying $p_T < 0.5$ GeV, relaxing the veto on unmatched clusters and extrapolating the number of events to zero unmatched clusters (the signal region) (see ref. [6] for details) .

Other sources of background are predicted to be negligible in the SRs. The uncertainty in the integrated luminosity is 1.9%, obtained using the LUCID-2 detector [13] for the primary luminosity measurements.

3 The prefit distributions

To verify the goodness of the Monte Carlo predictions compared to the data, the 2μ-CR events have been used. Figure 5 and Figure 6, show the leading muon p_T-distribution and the muon-pair invariant mass distributions in the 2μ-CR region. The data are compared to the predictions of STARLIGHT simulation improved at $p_T^{\mu} > 2GeV$ with events from the Madgraph Montecarlo which provides a better description of $\gamma\gamma \to \mu\mu(\gamma)$ processes. The agreement between prediction and data is fairly good, with a little contribution from non $\gamma\gamma \to \mu\mu$ events (including possible $\gamma\gamma \to \tau\tau$ events with both τ's decaying into muon + neutrinos).

Figure 7 shows for the 3 signal regions the prefit distributions of the muon-track p_T ($\mu 1T$-SR), left, 3track invariant mass distribution ($\mu 3T$-SR), central, and μe invariant mass distribution (μe-SR), right; Figure 7 shows the data compared to the signal and background contributions estimated with STARLIGHT+Madgraph Montecarlo and with a data-driven method for the $\mu\mu$ and photonuclear events, respectively.

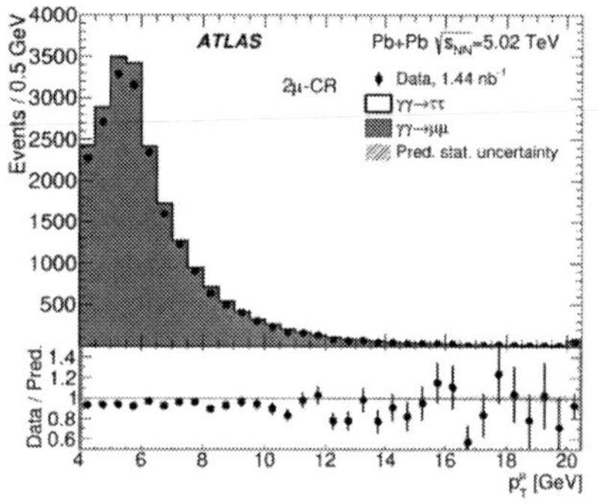

Figure 5. Leading muon p_T-distribution in 2μ-CR region [6].

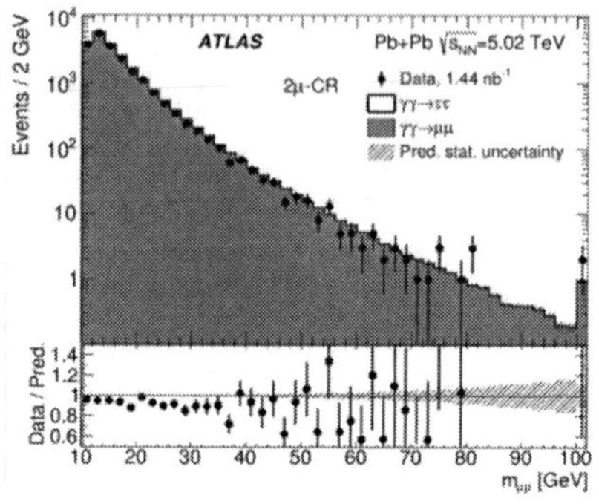

Figure 6. muon-muon invariant mass in 2μ-CR region [6].

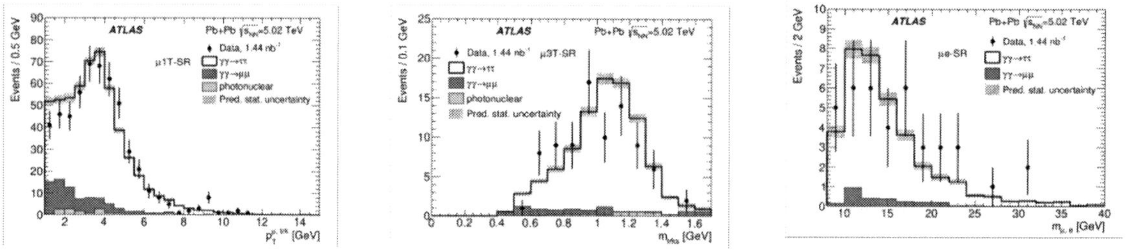

Figure 7. Signal region $\mu 1T$-SR: p_{T}-distribution of μ – *track* (left); Signal region $\mu 3T$-SR: 3-tracks invariant mass distribution (central); Signal region μe-SR: muon-electron invariant mass distribution (right) [6].

After applying the event selection, there are a total of 656 candidate events in the 3 signal regions with 80 and 16 estimated background events from $\mu\mu$ and photonuclear processes. The agreement between the predictions and the data is good for the three signal region plots.

4 Fitting the data

A profile likelihood fit to the muon-p_T distribution in the three SRs and 2μ-CR regions was used to extract the $\gamma\gamma \to \tau\tau$ signal strength: $\mu_{\tau\tau} = \frac{\sigma_{\tau\tau}^{data}}{\sigma_{\tau\tau}^{predicted}}$ and a_τ.

To measure a_τ, in particular, a fit is performed where a_τ is the only free parameter using the p_T^μ distribution in the three SRs and 2μ-CR; p_T^μ is chosen because of its high sensitivity to a_τ [7]. Simulated signal samples with various a_τ values are employed. In the nominal sample, a_τ is set to its SM value. Signal templates for alternative a_τ hypotheses, are obtained by modifying the $\gamma\tau\tau$ coupling in the Montecarlo-Lagrangian describing the process. Then the nominal sample is reweighted in three dimensions: differentially in $\tau\tau$ invariant mass, $\tau\tau$ rapidity, and rapidity difference between the two $\tau - leptons$, according to calculations from Ref. [7]. Note that the 3 kinematical parameters are sufficient to determine the p_T of the two τ's and, as a consequence, the p_T of the muon from the $\tau - decay$.

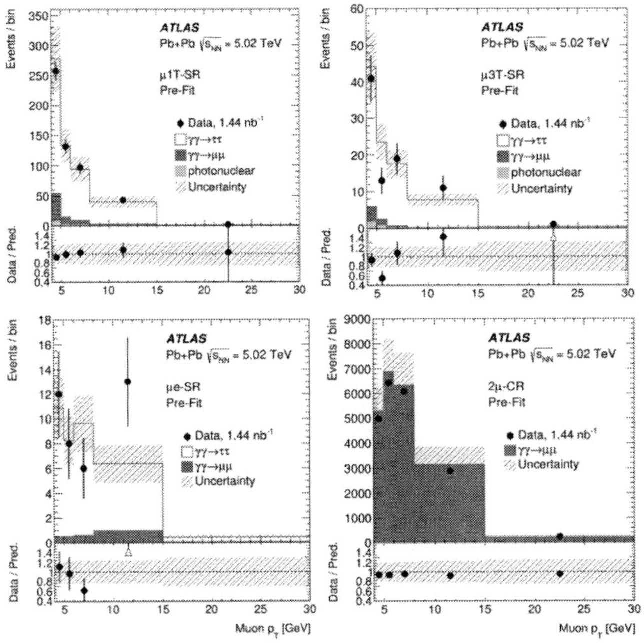

Figure 8. Prefit muon-p_T distributions for data and predictions, in the 3 signal regions and in the control region μ-CR. In the signal regions it is assumed the SM value of a_τ. Below each plot the ratio data/prediction is shown [6].

The main systematic uncertainties (in addition to the uncertainty in the photon flux) are: the muon-trigger efficiency, the muon/electron reconstruction/identification efficiency and calibration and the track reconstruction efficiency.

Figure 8 shows, for data and predictions, the muon-p_T distribution for the 3 signal regions and the control region 2μ-CR. The agreement between data and MC prediction is good; the largest systematic uncertainty in all distributions comes from the photon flux modelling which affects the signal yield by about 20%.

Figure 9 shows, for data and predictions, for the 3 signal regions and the control region 2μ-CR, the muon-p_T distributions. The results are obtained by a simultaneous fit using all regions. The inclusion of the 2μ-CR sample in the combined fit, strongly reduces the uncertainty originated by the photon flux modelling. The fit describes the data well. The evidence

 https://doi.org/10.1051/epjconf/202328201009

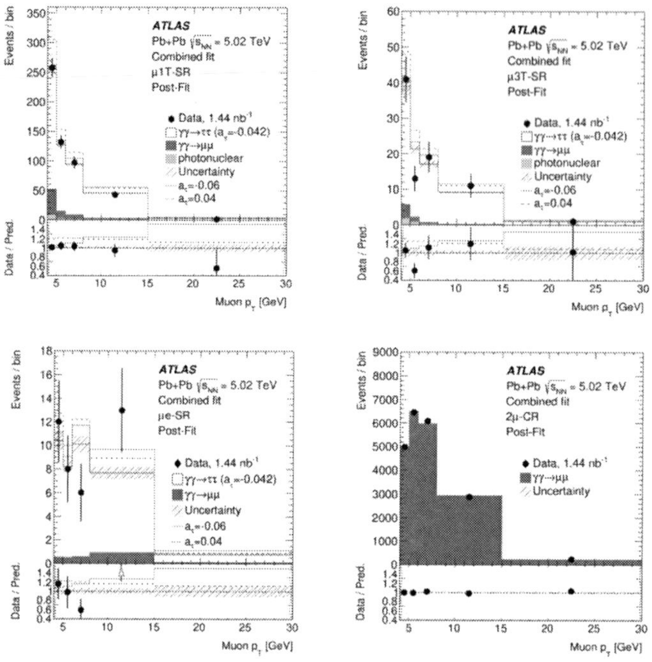

Figure 9. Muon p_T-distribution in μ1T-SR (top-left). Muon p_T-distribution in μ3T-SR (top-right). Muon p_T-distribution in μe-SR (bottom-left). Muon p_T-distribution in 2μ-CR (bottom-right). Black markers denote data and stacked histograms indicate the different components contributing to the regions. Post-fit distributions are shown with the signal contribution corresponding to the best-fit a_τ value (a_τ = 0.041). For comparison, signal contributions with alternative a_τ values are shown as solid red (a_τ =006) or dashed blue (a_τ= 0.04) lines. The bottom panel shows the ratio of the data to post-fit predictions. Vertical bars denote uncertainties from the finite number of data events. Hatched bands represent $\pm 1\sigma$ systematic uncertainties of the prediction with the constraints from the fit applied [6].

of the measurement of the process $\gamma\gamma \rightarrow \tau\tau$ is beyond 5 standard deviations. After the fit, the photon-flux uncertainty becomes subdominant and the luminosity uncertainty becomes negligible relative to other sources. The leading contributions to the total systematic uncertainty are: the estimation of the muon trigger efficiency, the τ-lepton decay modelling and the track reconstruction efficiency.

5 Signal strength and constraints on a_τ

The signal strength $\mu_{\tau\tau}$, defined as the ratio of the observed signal yield to the SM expectation assuming the SM value for a_τ , is measured to be: $\mu_{\tau\tau}$= 1.03 $^{+0.06}_{-0.05}$(tot) = 1.03$^{+0.05}_{-0.05}$(stat) $^{+0.03}_{-0.03}$(sys), largely dominated by the limited statistics. A total of 14 templates for different τ values are created to model the dependence of the $\mu - p_T$ distribution on a_τ in the three SRs. The combined-fit value of a_τ is a_τ = 0.041, with the corresponding 68% CL and 95% CL intervals being(-0.050,-0.029) and (-0.057, 0.024), respectively. The higher-than-expected observed yields lead to the highly asymmetric 95% CL interval. This arises from the nearly quadratic signal cross-section dependence on a_τ, caused by the interference of the SM and BSM amplitudes. The expected 95% CL interval is $-0.039 < a_\tau < 0.020$. The impact of systematic uncertainties on the final results is small relative to statistical uncertainties. Figure 10

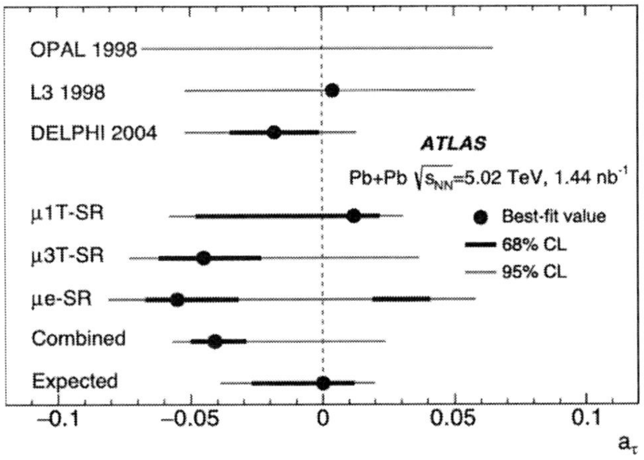

Figure 10. Measurements of a_τ from fits to individual signal regions (including the dimuon control region), and from the combined fit. These are compared with existing measurements from the OPAL, L3 and DELPHI experiments at LEP. A point denotes the best-fit a_τ value for each measurement if available, while thick black (thin magenta) lines show 68% CL (95% CL) intervals. The expected interval from the ATLAS combined fit is also shown [6].

shows the a_τ measurement alongside previous results obtained at LEP. The precision of this measurement is similar to the most precise single-experiment measurement by the DELPHI Collaboration.

6 Conclusions

In summary, τ-lepton pair production in ultraperipheral heavy-ion collisions, $Pb + Pb \rightarrow Pb(\gamma\gamma \rightarrow \tau\tau)Pb$, is observed by ATLAS with a significance exceeding 5 standard deviations in 1.44 nb^{-1} of $\sqrt{s_{NN}}$= 5.02 TeV data at the LHC. The observed event yield is compatible with that expected from the SM prediction within uncertainties. The events are used to set constraints on the a_τ-lepton anomalous magnetic moment, corresponding to $-0.057 < a_\tau < 0.024$ at 95% CL. The precision of the present experiment is limited by the statistical error: there is a large room for significant improvements in run3 and HL-LHC data taking periods.

References

[1] J. Schwinger, Phys. Rev. **73**, 416 (1948).
[2] D. Hanneke, S. Fogwell Hoogerheide and G. Gabrielse, Phys. Rev. A**83**, 052122 (2011).
[3] Muon g-2 collaboration, Phys. Rev. Lett. **126**, 141801 (2021).
[4] The DELPHI Collaboration, Eur. Phys. J. C**35**, 159 (2004).
[5] S. Eidelman and M. Passera, Mod. Phys. Lett. A**22**, 159 (2007).
[6] ATLAS Collaboration, arXiv:2204.13478 [hep-ex], accepted for publication on Phys. Rev. Lett.; https://atlas.web.cern.ch/Atlas/GROUPS/PHYSICS/PAPERS/STDM-2019-19/.
[7] M. Dyndal, M. Klusek-Gawenda, M. Schott and A. Szczurek, Phys. Lett. B **809**, 135682 (2020).
[8] E. Fermi, Z. Phys. **29**, 315 (1924); N. Cimento **2**, 143 (1925).
[9] ATLAS Collaboration, JINST **3**, S08003 (2008).
[10] ATLAS Collaboration, CERN-LHCC-2007-001 (2007); https://cds.cern.ch/record/1009649.
[11] S. R. Klein, J. Nystrand, J. Seger, Y. Gorbunov, and J. Butterworth, Comput. Phys. Commun. **212**, 258 (2017).
[12] N. Davidson, G. Nanava, T. Przedzinski, E. Richter-Was, and Z. Was, Comput. Phys. Commun. **183**, 821 (2012).
[13] G. Avoni et al., JINST **13**, P07017 (2018).

Precision measurements in the beta decay of ^{6}He

Mohamad Kanafani[1,*], *Xavier* Fléchard[1], *Oscar* Naviliat-Cuncic[1,2], *Gyang Davou* Chung[1], *Sylvain* Leblond[3], *Etienne* Liénard[1], *Xavier* Mougeot[3], *Gilles* Quéméner[1], *Adriana* Simancas Di Filippo[1,4], and *Jean-Charles* Thomas[5]

[1]Université de Caen Normandie, ENSICAEN, CNRS/IN2P3, LPC Caen UMR6534, F-14000 Caen, France
[2]Facility for Rare Isotope Beams and Department of Physics and Astronomy, Michigan State University, East Lansing 48824 MI, USA
[3]Université Paris-Saclay, CEA, List, Laboratoire National Henri Becquerel (LNE-LNHB), F-91120 Palaiseau, France
[4]Deutsches Elektronen-Synchrotron DESY, Notkestraße 85, 22607 Hamburg, Germany
[5]GANIL, CEA/DRF-CNRS/IN2P3, Bd Henri Becquerel, 14076 Caen, France

Abstract. We report here about the ongoing data analysis of an experiment performed at GANIL with a 25 keV $^{6}He^{+}$ beam to determine the Fierz interference term from the β particles energy spectrum.

1 Introduction

The ^{6}He beta decay played a fundamental role in establishing the V-A character of the weak interaction [1, 2]. The fact that ^{6}He decays into ^{6}Li by a pure Gamow-Teller transition made it attractive for searches of physics beyond the standard model. The Fierz interference term is one of the coefficients that can be used to probe new physics, since it is linearly dependent on exotic tensor and scalar couplings. The Fierz term can be accessed experimentally by high precision measurements of the shape of the β-energy spectrum.

2 Experimental setup

The apparatus is described in details in Ref [3]. An experiment with a low-energy beam of $^{6}He^{+}$ ions was performed at the Grand Accélérateur National d'Ions Lourds (GANIL), Caen. The $^{6}He^{+}$ ions were guided at 25 keV towards the surface of a fixed detector "det 1". A second identical and movable detector "det 2" is used to enclose the ions implantation region and to achieve a 4π solid angle as shown in Fig 1. The motion of "det 2" is accurately synchronized with the beam implantation and the data acquisition. The period of the cycles is chosen such as to determine with a sufficient precision the surrounding background. The detector coverage ensures the full collection of all β particles emitted by the implanted $^{6}He^{+}$ ions, eliminating thus any energy loss due to backscattering.

The detectors "det 1" and "det 2" are formed each of a cylindrical $YAlO_3$ Ce-doped inorganic scintillator (YAP) surrounded by an EJ-204 plastic scintillator. The two scintillators are mounted in a phoswich configuration in which both of them are read out by one single

*Corresponding author: kanafani@lpccaen.in2p3.fr

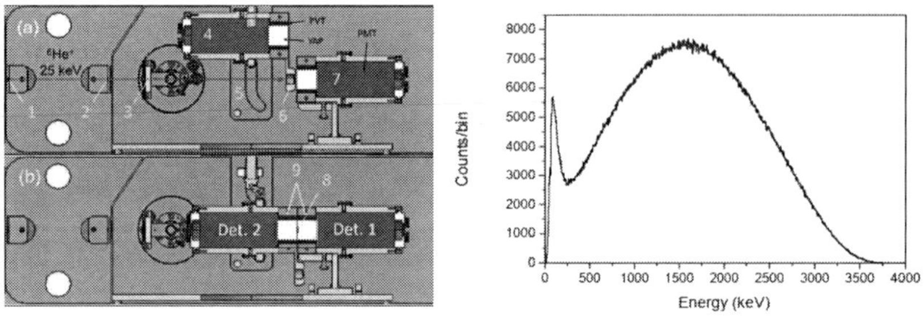

Figure 1. Left panel: sectional view of the experimental setup during beam implantation (a) and data taking (b). The labels on panel (a) are: 1 and 2– the two Ø6 mm collimators in the first section of the chamber; 3– a movable Si detector; 4 and 5– the moving detector and its mechanical guide; 6– the third Ø4 mm collimator; 7– the fixed detector. The green arrow indicates the $^{6}He^{+}$ beam. On panel (b), label 8 indicates the implantation region and 9 the two ^{241}Am calibration sources [3]. Right panel: experimental β-energy spectrum for one run of two hours duration.

photomultiplier tube (Fig 1). A 5-kBq ^{241}Am source is mounted on each of the detectors as illustrated in Fig 1. The 59.54 keV γ rays from the ^{241}Am were used as a reference to monitor gain and baseline variations during the experiment.

Each event is labeled with a time stamp and an energy charge integration, allowing a control over systematic effects in the offline analysis. Five sets of runs were taken with different experimental conditions, to study systematic and background effects [3].

3 Background investigation

The experimental β-energy spectra showed the expected continuous spectrum extending up to 3.5 MeV, the endpoint energy of ^{6}He decay, alongside with an unexpected contribution peaked at 0.1 MeV (Fig 1). This peak was also present in the background runs, where the ^{6}He beam was implanted on the collimator fixed to the moving detector, and not on the YAP (Fig 2). This peak was identified to be caused by β particles from the $^{6}He^{+}$ ions, interacting with the material of the collimator itself and generating Bremsstrahlung photons that are detected by the two detectors.

The experimental geometry with the two detectors and the collimator attached to "det 2" was built in GEANT4 (Fig 2), to validate the origin of this peak at low energy. Events were generated using the phase space of the ^{6}He decay, on the inner surface of the collimator as shown in Fig 2. The deposited energy spectrum inside the two YAP scintillators obtained by simulation was found to match the experimental spectrum up to 1 MeV (Fig 2, left).

A wide peak between 1.5 and 3 MeV, which is not reproduced by the former simulation, was also observed in the background data (Fig 2, left). This peak was attributed to electrons from ^{6}He decay, going through a hole in the lower part of the detector, which leads to the YAP scintillator of the movable detector through the plastic scintillator. Another simulation was then performed where the source of electrons was set on the outer surface of the collimator. The simulated spectrum of the deposited energy inside the YAP scintillators showed the appearance of the same distribution along with the Bremsstrahlung peak (Fig 2, center), confirming thereby the origin of these events.

For the β spectrum shape analysis, the presence of these two intruders within the β-energy spectrum will be suppressed by using data taken during the background runs.

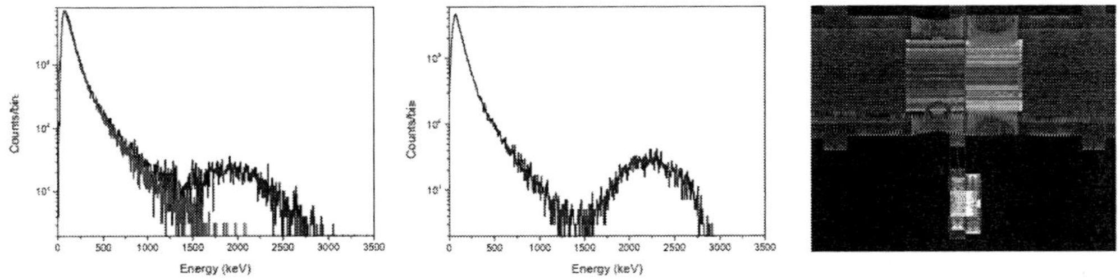

Figure 2. Left panel: the blue histogram is the experimental energy spectrum for one run where the ^{6}He was implanted on the collimator. The red histogram is the deposited energy spectrum obtained with simulation, normalized to the experimental data. Center panel: the deposited energy spectrum obtained with a simulation where the source of electrons is placed on the outer surface of the collimator. Right panel: the two detectors with the collimator built in GEANT4. The orange and red lines represent respectively the position of the source of events on the inner and outer surface of the collimator. The red circle shows the position of the hole through which the electrons of the decay on the collimator were able to access the YAP scintillator.

4 β-energy spectrum fit function

The energy distribution of the β particles emitted from ^{6}He decay can be written as:

$$N(W) \propto pW(W - W_0)^2 \left(\alpha_0 + \alpha_{-1} \cdot \frac{1}{W} + \alpha_1 \cdot W + \alpha_2 \cdot W^2\right) \tag{1}$$

where p, W and W_0 are respectively the momentum, total energy and the endpoint energy of the emitted electrons in unit of the electron mass. α_{-1}, α_0, α_1 and α_2 are coefficients that represent all the relevant corrections to the shape of the spectrum (the Fierz term is included in α_{-1}). It can be useful to have these coefficients fixed with different values or set as free parameters in the fitting procedure. However, this function cannot be used directly to fit the experimental β-energy spectrum, since it does not account for the energy loss due to Bremsstrahlung energy escape. In the following, we introduce a method used to account for Bremsstrahlung escape within an analytical model.

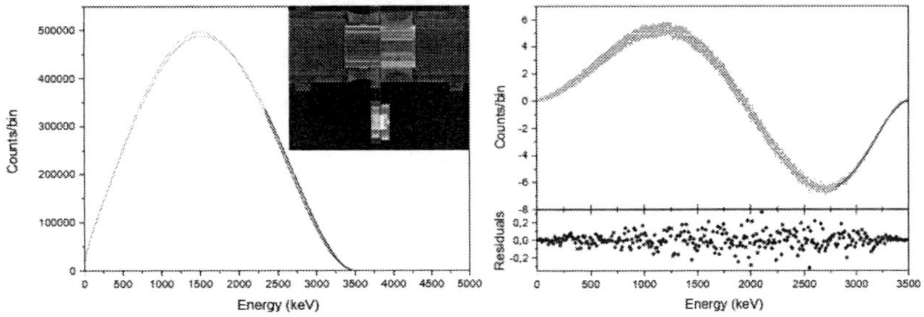

Figure 3. Left panel: the generated energy spectrum with Eq. (2) (in blue), and the deposited energy within the two YAP scintillators obtained by GEANT4 simulation (in red). Insert: the simulation geometry with the source of events placed between the two YAP scintillators (the red spot). Right panel: on top, the histogram of the normalized difference between the generated and deposited energy histograms (in blue). The fit of this histogram with a polynomial function (in red). On bottom, the residuals of the fit.

As a first step, four distributions of events were generated following the functions:

$$g_i(W) = pW(W - W_0)^2 \cdot W^i \tag{2}$$

where $i = -1, 0, 1, 2$. For each $g_i(W)$ distribution, the effect of the Bremsstrahlung escape was estimated using GEANT4. The energy spectra of the generated and the deposited energies were built for each of the four sets. The histograms of the normalized difference between the generated and deposited energy spectra were then plotted and fitted with a polynomial function $f_i(W)$ (Fig 3). These functions represent the effect of the Bremsstrahlung escape on each term of the energy spectrum.

The deposited energy spectrum resulting from the decay function of Eq. (1) and accounting for Bremsstrahlung escape can be expressed or fitted with:

$$F(W) = \alpha \sum_{i=-1}^{2} \alpha_i \left[g_i(W) + f_i(W) \right] \tag{3}$$

where α is a normalization factor and $f_i(W)$ are the polynomial functions that describe the effect of Bremsstrahlung energy escape for each term of the β-energy spectrum. These functions were tested with several statistically independent data sets generated in GEANT4 with Eq. (1) for several values of α_{-1}, α_0, α_1 and α_2. The deposited energy spectra were built afterwards and fitted with Eq. (3), where α_0, α_1 and α_2 were fixed parameters, while α and α_{-1} were free parameters.

The results of the fits were all statistically consistent with the values of α_1 that were initially used to generate the spectra. The standard residuals of the fit function, which are distributed around zero within $\pm 2\sigma$, also proved the validity of the fit function (Fig 3).

5 Summary

To summarize, the analysis of the spectrum shape of ^{6}He has been briefly introduced. Two background components present within the experimental β spectrum have been identified. They will be dealt with using background runs, during which $^6He^+$ ions were blocked from reaching the detector surface. The effect of the Bremsstrahlung energy escape on the shape of the β-energy spectrum was presented alongside how we plan to account for this effect with the addition of a polynomial analytical function which will be used later on to fit the experimental data.

This project was supported in part by the French Agence Nationale de la Recherche under grant ANR-20-CE31-0007-01 (bSTILED).

References

[1] C. H. Johnson, F. Pleasonton, and T. A. Carlson, Phys. Rev. **132**, 1149 (1963).
[2] J. B. Vise and B. M. Rustad, Phys. Rev. **132**, 2573 (1963).
[3] M. Kanafani, X. Fléchard, O. Naviliat-Cuncic, G. D. Chung, S. Leblond, E. Liénard, X. Mougeot, G. Quéméner, A. Simancas Di Filippo, and J.-C. Thomas, Phys. Rev. C **106**, 045502 (2022).

Neutrino mass measurement and sterile neutrinos search with the KATRIN experiment

Anthony Onillon[1,*] on behalf of the KATRIN collaboration

[1]Physik-department, Technische Universität München, D-85748 Garching, Germany

Abstract.
In this proceeding, we report on the latest results of the KATRIN experiment on the effective electron anti-neutrino mass and the search for sterile neutrinos. KATRIN performed a high resolution and high statistics spectroscopy measurement of the tritium β-decay spectrum around its endpoint energy. The analysis of ~1265 h of data acquired during the first two measurement campaigns of 2019 led to the most stringent limit on the neutrino mass with $m_\nu < 0.8$ eV (90% CL). We also report on the improved KATRIN exclusion limits for the eV- and keV-scale sterile neutrinos search. Finally, the future of KATRIN with the TRISTAN detector aiming at further improve the laboratory-based sensitivity to keV-scale sterile neutrino is discussed.

1 Introduction

Since the existence of neutrinos was suggested almost a hundred years ago by W. Pauli, neutrinos have benefited from an intensive research program aimed at measuring their properties. While it is now experimentally established that there are only three flavors of active neutrinos with different masses [1], the question of the mass hierarchy, but also of the absolute mass of the neutrinos and of the neutrinos nature itself remained unanswered [2]. In particular, the neutrino mass which cannot be probed by oscillation experiments, is of fundamental importance in cosmology due to its role in the evolution of the universe but also in particle physics for its role in the models beyond the SM [3]. Different types of measurement can be used to probe the neutrino mass: cosmological observation, search for neutrinoless double beta decay and kinematic study of beta decay. In this paper, we focus on the latter.

2 The KATRIN experiment

The Karlsruhe Tritium Neutrino (KATRIN) experiment [4, 5] is designed to measure the mass of the electron anti-neutrino with a sensitivity of 0.2 eV at 90% confidence using the kinematics of the tritium β-decay [6]:

$$T_2 \rightarrow {}^3\mathrm{HeT}^+ + e^- + \bar{\nu}_e. \quad (1)$$

The low Q-value of 18.6 keV of the tritium β-decay combined with its short half-life of $T_{1/2} = 12.3$ yr and its super-allowed nature make the tritium a well-suited isotope for absolute neutrino mass measurement. The neutrino mass can be measured by looking for a

*e-mail: anthony.onillon@tum.de

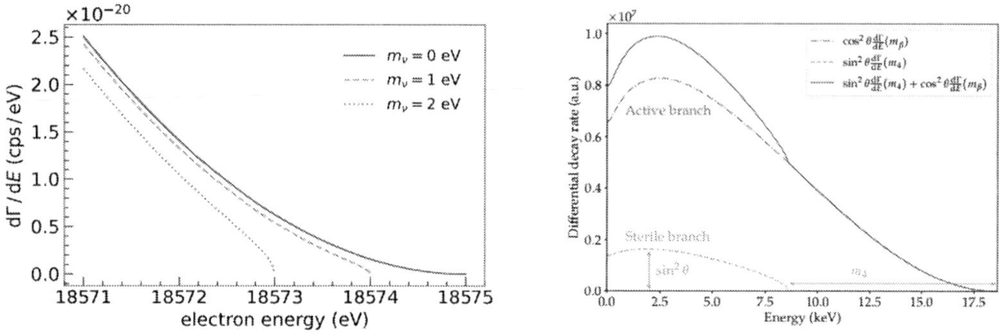

Figure 1. (Left) Illustration of the shape distortion induced by the neutrino mass on the tritium β-decay spectrum. (Right) Illustration of the sterile neutrino signature in a differential tritium β-decay spectrum. For the sake of clarity, a sterile neutrino with a mass of 10 keV and an unrealistic large mixing angle is considered.

reduction of the measured maximal electron energy as well as to a slight spectral shape distortion compared to the theoretical prediction. In the presence of mixing and for neutrinos with small mass differences the distortion of the beta spectrum is given by [6]:

$$\frac{dN}{dE} \simeq R(E) \sum_i |U_{ei}|^2 \sqrt{(E_0 - E)^2 - m_\nu^2} \tag{2}$$

where $E_0 = 18.575$ keV [7] is the endpoint energy of the tritium β-decay and m_ν is the effective electron anti-neutrino mass defined as the incoherent sum of neutrino mass eigenstates: $m_\nu^2 = \sum |U_{ei}|^2 m_i^2$, where $|U_{ei}|^2$ are the elements of the Pontecorvo-Maki-Nakagawa-Sakata matrix describing the mixing of the neutrino states i. $R(E)$ contains all m_ν-independent factors. The impact of the neutrino mass on the β-decay spectrum of the tritium is illustrated in figure 1. Because of the small size of the distortion and the very small number of events expected around the endpoint energy, such measurement represents an experimental challenge. A high luminosity source and a high resolution on the electron-volt scale are thus required to address these constraints.

3 Experimental setup

The KATRIN experiment consists in a 70-m-long beamline located at the Karlsruhe Institute of Technology (see figure 2). The beamline is divided into the tritium-containing source and transport section (STS) and the tritium-free section containing the spectrometers and detector system (SDS). Electrons are produced in a 10 m long windowless gaseous tritium source (WGTS) in which a highly purified molecular tritium gas is continuously injected. The gas diffuses towards both ends of the tube where it is pumped out and fed back to the tritium loop system. The WGTS is designed to provide an extremely high activity of up to 10^{11} Bq. It is cooled at 30 K to minimize the thermal gas motion and measured stable at the 0.2% level [8]. Electrons are magnetically guided from the source to a pre- and a main spectrometer through a transport section. Between the source and the spectrometers, the differential tritium pressure is reduced by more than 14 orders of magnitude thanks to several pumping sections. The two spectrometers rely on the MAC-E principle and act as high-pass filters that allow only electrons with kinetic energies larger than the applied retarding energy qU to be transmitted. By applying a magnetic field gradually reduced toward the

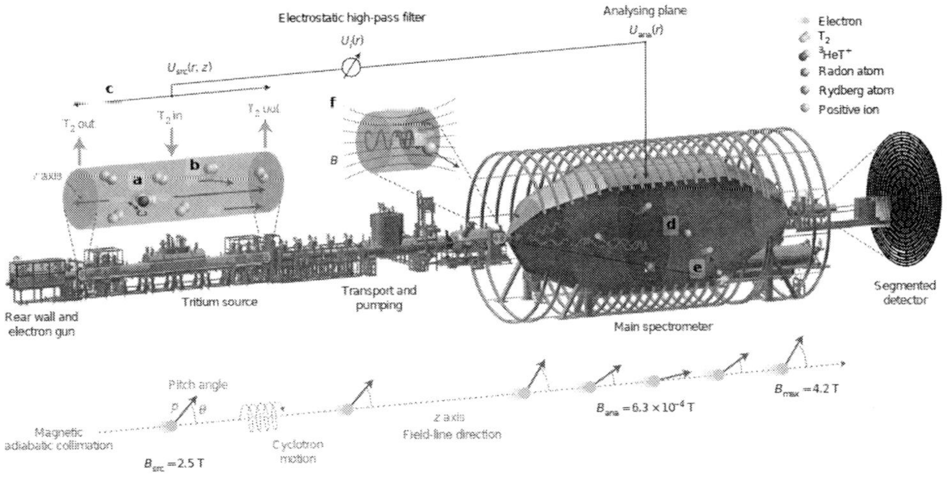

Figure 2. Illustration of the KATRIN beamline [8]. The 70-m-long beamline is composed of a rear wall section, a tritium source, a transport and pumping section, a pre- and a main spectrometer and a segmented focal plane detector.

analysing plane, electrons are adiabatically guided through the spectrometers. The MAC-E filter technology has the advantages of combining a large angle acceptance of 51° with an energy resolution at the eV-scale. The pre-spectrometer is used to reduce the flux of electrons into the main spectrometer by transmitting only electrons with an energy above 10 keV. Electrons transmitted through the main spectrometer are finally counted at the focal plane detector (FPD) as a function of the retarding potential . At the other end of the setup is located the rear section. It contains a gold-plated rear wall absorbing the non-transmitted electrons and defines the electric ground potential relative to the high voltage of the main spectrometer. It also contains an electron gun for calibration measurements.

4 Neutrino mass measurement

The integral beta spectrum is acquired by repetitively recording the detector count rate at different retarding potential. A complete sequence of such points being termed a scan. The scans are performed in an interval of 40 eV below the endpoint and 135 eV above. About 40 scans with dwell times at each point chosen to optimize the neutrino-mass sensitivity are used. The scans above the endpoint are used to measure the background. The most recently published results of KATRIN use the data acquired in 2019 during the first two measurement campaigns KNM1 and KNM2 [8]. A total of 1.48×10^6 (3.68×10^6) β-electrons were collected in the range of interest during a 521.7 h (743.7 h) period of data taking for KNM1 (KNM2).

The neutrino mass is inferred by comparing the measured integral spectrum to the expected one. The expectation is obtained by coupling a very accurate theoretical prediction of the tritium beta spectrum [9] with the experimental response obtained through calibration. This expectation is fitted to the data with four free parameters: the signal amplitude, the endpoint, the background rate and the neutrino mass. Systematic uncertainties are propagated using the covariance matrix approach. A simultaneous fit of both datasets yields $m_\nu^2 = (0.1 \pm 0.3)\ eV^2$ and a corresponding upper limit of $m_\nu < 0.8$ eV at 90% CL. This results supersedes

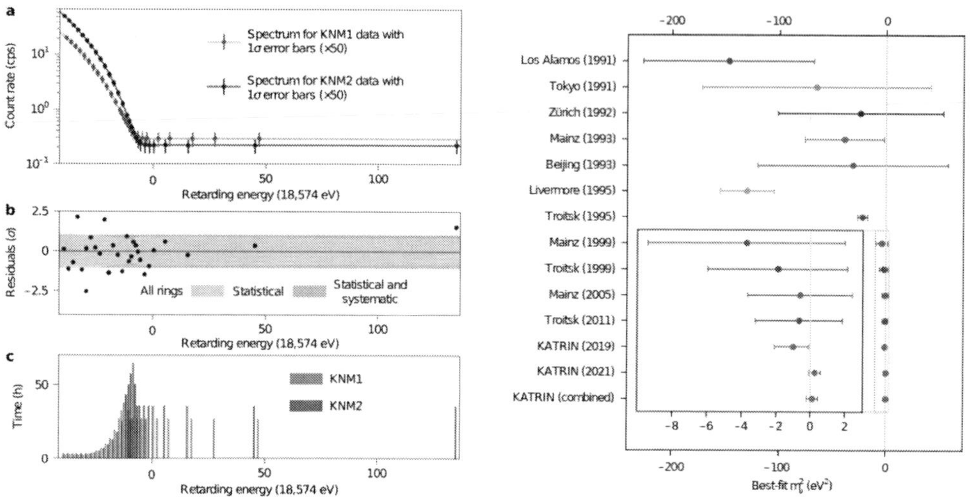

Figure 3. (Right): Measurement rate at each retarding energy for the first two KATRIN campaigns KNM1 and KNM2 [8]. (a) Data points for the count rate over all the detector rings. Error bars represent the statistical uncertainties multiplied by a factor of 50. Best-fit models for each campaign are represented by the colored lines. A reduced background rate, higher signal strength and overall higher statistics can be observed for the second campaign. (b) Normalized residuals for the fit to the data from KNM2. The shaded areas represent the statistical and total uncertainties. (c) Measurement time distribution for the two campaigns. (Left): Comparison of KATRIN results for the neutrino mass with past measurements [8]. The error bars represent the combined statistical and systematic uncertainties.

the previous KATRIN limit of $m_\nu < 1.1$ eV at 90% CL [10] and is to date the most stringent experimental constraint on the neutrino mass. The neutrino-mass result is dominated by statistical uncertainty and the largest systematic uncertainties are related to the background properties and the source electric potential. Correlated systematic uncertainties between both campaigns are negligible as both datasets are dominated by statistics. The data and best-fit models for the first two KATRIN campaigns as well as a comparison of KATRIN neutrino mass results with past measurements are presented in figure 3.

5 Sterile neutrino search

If it is well established that there are only three active neutrinos [1], the existence of one or more sterile neutrino that could only interact through gravitation has been proposed to explain some unexpected experimental results. On one hand, a light sterile neutrino [11] with a eV-scale mass could alleviate the anomalies observed in short-baseline neutrino oscillation experiments such as the gallium anomaly (GA) characterised by a deficit of ν_e from ^{37}Ar and ^{51}Cr electron capture decays [12–14] and the reactor antineutrinos anomaly (RAA) characterised by a deficit of $\bar{\nu}_e$ from nuclear reactors [15]. On the other hand, a sterile neutrino with a keV mass scale would be a viable candidate to the dark matter and could help to explain the observed pulsar velocities [3]. The existence of eV and keV neutrinos has not yet been ruled out and KATRIN appear to be very suitable as a complementary laboratory based experiment to investigate this hypothesis.

5.1 eV-sterile neutrino search

Beta decay can be used to investigate the existence of eV-sterile neutrinos. The signal signature would consist in a kink-like spectral distortion at $E = E_0 - m_4$, where E_0 is the endpoint energy of the tritium and m_4 the mass of the fourth neutrino state. The distortion of the β-decay spectrum can be expressed as:

$$R_\beta(E) = \cos^2\theta \cdot R_\beta(E, m_\nu^2) + \sin^2\theta \cdot R_\beta(E, m_4^2), \tag{3}$$

where $R_\beta(E, m_\nu^2)$ and $R_\beta(E, m_4^2)$ correspond respectively to the active and sterile decay branches and θ is the active-sterile neutrino mixing angle. R_β denotes here the tritium beta spectrum. The size of the signal is determined by the admixture $\sin^2\theta$ of the new mass eigenstate to the electron neutrino flavor. An illustration of the sterile neutrino imprint in a differential beta spectrum is illustrated in figure 1.

Using the same datasets as for the active neutrino mass, KATRIN investigated the existence of a fourth neutrino mass eigenstate [16]. The same energy range of [E_0 - 40 eV, E_0 + 135 eV] was used. The sterile fit was achieved by inferring the two parameters of interest m_4^2 and $|U_{e4}|^2$ ($|U_{e4}|^2 = \sin^2\theta$) in addition to the original fit parameters used to measure the neutrino mass. Similarly, systematic uncertainties were propagated using the covariance matrix approach. This analysis is sensitive to the fourth neutrino mass eigenstate $m_4^2 \lesssim 1600$ eV^2 and active-to-sterile mixing $|U_{e4}|^2 \gtrsim 6 \times 10^{-3}$ in the $3\nu + 1$ framework. The measurement is statistics dominated over the whole analysed energy range with $\sigma_{stat}^2/\sigma_{tot}^2 > 0.5$ for all m_4^2. No significant eV-sterile neutrino signal was observed and an improved exclusion limit was reported (see figure 4). More stringent constraints than both Mainz and Troitsk experiments for $m_{41}^2 \lessapprox 300$ eV^2 were obtained and large δm_{41}^2 solutions of the RAA and of BEST+GA anomalies are now excluded. Notably, this new result disfavors the Neutrino-4 hint of a signal for $\sin^2(2\theta_{ee}) \gtrsim 0.4$ [17] and improves the exclusion bounds set by short-baseline oscillation experiments for $m_{41}^2 \gtrapprox 10$ eV^2.

5.2 keV-sterile neutrino search

Like eV-sterile neutrinos, keV-sterile neutrinos can be investigated by searching for a kink-like distortion in the tritium β-decay spectrum. Because of the unconstrained mass, such sterile signature would lie far away from the endpoint and would require a measurement of the spectral shape deep into the spectrum. Thanks to its high-luminosity source and the high endpoint value of the tritium, KATRIN exhibits a high potential to search for keV sterile neutrinos with mass $m_4 < E_0 \simeq 18.6$ keV. However, scanning the entire spectrum with the full source activity would lead to an increase of the signal rate by several orders of magnitude compared to the standard KATRIN measurement. The current detector system is not designed to handle such high count rates. In 2018, KATRIN acquired its first data during a commissioning campaign aiming at demonstrating the stability of the operation of the experiment. This campaign was realized with a reduced isotopic abundance of 0.5% of the tritium in the deuterium carrier gas. Thanks to this reduced activity and a higher retarding potential, this data set provided a unique opportunity to search for sterile neutrinos in the 0.01 - 1.6 keV mass range [18]. No keV-sterile neutrino signal was observed and KATRIN reported exclusion limits competitive with previous laboratory-based searches. In particular, an active-sterile mixing amplitude of $\sin^2\theta < 5 \times 10^{-4}$ for a sterile neutrino mass of $m_4 = 300$ eV was excluded and the currently leading laboratory-based bounds in a mass range of 0.1 keV $< m_4 <$ 1.0 keV were improved. As well as for the neutrino mass and the eV-sterile search, this result is dominated by statistical uncertainties.

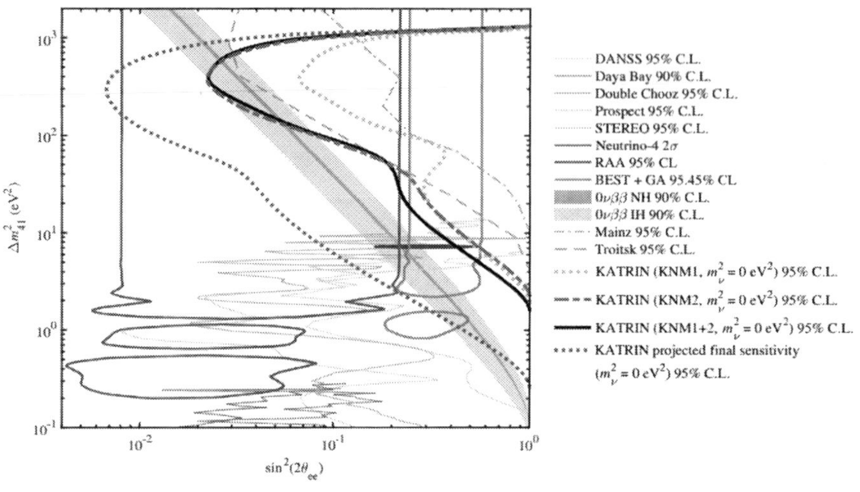

Figure 4. eV-scale sterile neutrino exclusion contours at 95% C.L. for the first and second campaign of KATRIN [16]. The combined results of the two campaigns as well the projected final sensitivity are also shown. These exclusions are compared to the most recent results from reactor, gallium, beta decay and double beta decay experiments. References for the different experimental results used in this plot can be found in [16].

This result represents an important milestone for the experiment as it demonstrates the feasibility of using KATRIN to search for keV-sterile neutrinos. In order to handle the high counting rate that would arise by measuring the full tritium spectrum with a higher source activity, a new detector system is currently being developed. The TRISTAN (TRitium Investigation on STerile to Active Neutrino mixing) detector aims at using the full KATRIN source strength and reach a statistical sensitivity at the level of $\sin^2\theta \sim 10^{-6}$ in the $\sim$7 keV region. It was demonstrated that the observation of a kink-like structure requires an energy resolution in the order of 300 eV at 30 keV [19]. Silicon drift detectors (SDD) meet such requirement while also exhibiting low energy threshold (< 2 keV) and being able to handle high counting rate ($\sim 10^5$ cps for 3 mm pixel [19]). The development of the TRISTAN detector system follows a staged approach. First, a 7-pixels detector prototype was used to investigate the general detector system properties. The system was then scaled up with modules of 47 and 166 pixels in a monolithic chips. Extensive tests and characterisation of the different detector systems are conducted. This includes among others quantities the energy resolution, the linearity and the electronic noise. For this purpose, a large variety of calibration sources such as X-ray lines from ^{55}Fe, conversion electron lines from ^{83m}Kr, laser and electron beam have been used. These extensive testing have already demonstrated the good performance of the detectors and their ability to fullfill requirements for TRISTAN [19–21]. Ultimately, TRISTAN will consist in a focal plane array composed of respectively 9 and 21 detector modules for the first and the second phase of the keV-sterile search. Each module is composed of 166 pixels for a total of respectively $\sim$1500 and $\sim$3500 pixels for the first and second phase. An illustration of the TRISTAN detector for the second phase is presented in figure 5.

In addition to the detector development, the modelling of the expected spectrum at the detector over the full energy range is also an ongoing effort. The model includes non-trivial effects such as electrons backscattered at the rear wall and electrons backscattered and back-reflected at the detector due to the beamline magnetic fields. This model is used to investigate the impact of the different systematic effects and their associated uncertainties on the keV-

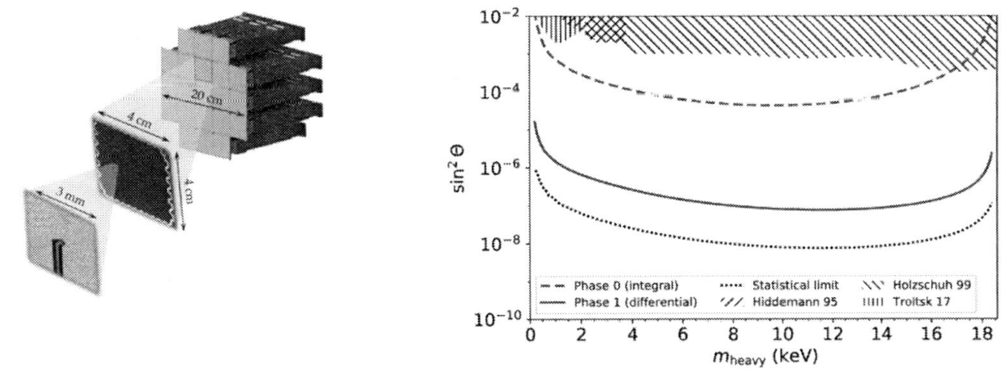

Figure 5. (Left) TRISTAN detector design for the second phase of the keV-sterile search. The full detector will consist of an array of 21 identical modules each consisting of 166 pixels. (Right) Expected 95% CL statistical sensitivity of the KATRIN experiment to keV-scale sterile neutrinos for different stages with the TRISTAN detector [19]. Phase-0 (dashed red line): integral measurement with a total statistics of 6×10^{11} electrons. This could be achieved with a 7 d measurement campaign at a reduced signal rate of 10^6 cps. This rate can be obtained with a reduced amount of tritium. Phase-1 (solid red line): differential measurement with a total statistics of 10^{16} electrons. This could be achieved with a 3 years measurement campaign and a signal rate of 10^8 cps. This signal rate can be obtained with a 100-fold reduced tritium column density (with respect to nominal KATRIN operation). The statistical limit (dashed black line) for a differential measurement with a total statistics of 10^{18} electrons is also reported. This limit represent the most optimistic scenario for which a measurement of 3 years with the full source strength of the KATRIN experiment can be achieved. The grey dashed area depicts the current laboratory-based limits [22–24].

sterile sensitivity. As of now, systematic effects are estimated to reduce by at least one order of magnitude the statistical limit. An illustration of KATRIN sensitivity expected for the keV-sterile search is presented in figure 5.

6 Conclusion

In early 2022, the KATRIN experiment reported its results on the neutrino mass and the search for eV- and keV sterile neutrinos. The analyses were performed on the data taken during the first two measurement campaigns conducted in 2019 for which a total of $\sim 6\times10^6$ β-electrons have been collected. The combination of the first two campaigns results gives an improved uper limit on the neutrino mass of $m_\nu < 0.8$ eV (90% CL). This result constitutes the best neutrino-mass sensitivity for direct neutrino-mass measurement. KATRIN also reported an updated analysis of its light sterile neutrino search. This analysis is sensitive to the fourth neutrino mass eigenstate $m_4^2 \lesssim 1600$ eV^2 and active-to-sterile mixing $|U_{e4}|^2 \gtrsim 6\times10^{-3}$ in the $3\nu + 1$ framework. No significant eV-sterile neutrino signal was observed and an improved exclusion limit was reported. Notably, this improved result disfavors the Neutrino-4 hint of a signal for $\sin^2(2\theta_{ee}) \gtrsim 0.4$. These two analysis are dominated by statistical uncertainties. KATRIN is currently analysing the data of three additional measurement campaigns corresponding to a statistical increase of ~5 compared to the analysis reported here. By the end of the data taking, a total statistic of about $\sim 70\times10^6$ electrons is expected for a target sensitivity on the neutrino mass of $m_\nu < 0.2 - 0.3$ eV (90% CL).

Moreover, KATRIN reported the results of its first keV-scale sterile-neutrino search using first tritium data acquired in the commissioning run. No significant keV-sterile neutrino signal was observed and exclusion limits were reported. This result improves the current laboratory limits on the active-to-sterile mixing amplitude in a mass range of 0.1 keV < m_4 < 1.0 keV by up to an order of magnitude. After completion of the neutrino-mass campaigns in 2025, KATRIN is planning to further improve the sensitivity to keV-scale sterile neutrinos. For this purpose, a novel detector system relying on the SDD technology is under development.

We acknowledge the support of Helmholtz Association, Ministry for Education and Research BMBF (5A17PDA, 05A17PM3, 05A17PX3, 05A17VK2, and 05A17WO3), Helmholtz Alliance for Astroparticle Physics (HAP), Helmholtz Young Investigator Group (VH-NG-1055), and Deutsche Forschungsgemeinschaft DFG (Research Training Groups GRK 1694 and GRK 2149, and Graduate School GSC 1085 - KSETA) in Germany; Ministry of Education, Youth and Sport (CANAM-LM2011019, LTT19005) in the Czech Republic; and the United States Department of Energy through grants DE-FG02-97ER41020, DE-FG02-94ER40818, DE-SC0004036, DE-FG02-97ER41033, DE-FG02-97ER41041, DE-AC02-05CH11231, DE-SC0011091, and DE-SC0019304, and the National Energy Research Scientific Computing Center.

References

[1] S. Schael et al., Phys. Rep. **427**, 257 (2006)
[2] R. Cahn et al., arXiv:1307.5487 (2013)
[3] R. Adhikari et al., JCAP **025**, 01 (2017)
[4] J. Angrik et al., FZKA scientific report, 1 , https://publikationen.bibliothek.kit.edu/270060419 (2005)
[5] M. Aker et al., J. Instrum. **16**, T08015 (2021)
[6] R.L. Workman et al., Prog. Theor. Exp. Phys. 2022, 083C01 (2022)
[7] E. G. Myers et al., Phys. Rev. Lett. **114**, 013003 (2015)
[8] The KATRIN Collaboration, Nat. Phys. **18**, 160–166 (2022)
[9] M. Kleesiek et al., Eur. Phys. J. C **79**, 204 (2019)
[10] M. Aker et al., Phys. Rev. Lett. **123**, 221802 (2019)
[11] K. N. Abazajian et al., arXiv:1204.5379 (2012)
[12] W. Hampe et al., Phys. Lett. B **420**, 114 (1998)
[13] J. Abdurashitov et al. Phys. Rev. C **80**, 015807 (2009)
[14] V. V. Barinov et al., Phys. Rev. Lett. **128**, 232501 (2022)
[15] G. Mention et al., Phys. Rev. D **83**, 073006 (2011)
[16] M. Aker et al., Phys. Rev. D **105**, 072004 (2022)
[17] A. P. Serebrov et al., Pis'ma Zh. Eksp. Teor. Fiz. **109**, 209 (2019)
[18] M. Aker et al., to be published, arXiv:2207.06337 (2022)
[19] S. Mertens et al., J. Phys. G **46**, 065203 (2019)
[20] S. Mertens et al., J. Phys. G **48**, 1, (2020)
[21] M. Biassoni et al., Eur. Phys. J. Plus **136**, 1, (2021)
[22] K.-H. Hiddemann et al., J. Phys. G **21**, 5, (1995)
[23] E. Holzschuh et al., Phys. Lett. B **451**, (1999)
[24] J. N. Abdurashitov et al., Pisma Zh. Eksp. Teor. Fiz. **105**, 12, (2017)

Future facilities at PSI, the High-Intensity Muon Beams (HIMB) project

Giovanni Dal Maso[1,2,*], *Alexander* Gabard[1], *Malek* Haj Tahar[1], *Peter-Raymond* Kettle[1], *Klaus* Kirch[1,2], *Daniela* Kiselev[1], *Alexander* Koschik[1], *Andreas* Knecht[1], *Daniel* Laube[1], *Angela* Papa[1,3,4], *Thomas* Rauber[1], *Davide* Reggiani[1], *Rebecca* Riccioli[1], *Jochem* Snuverink[1], *Vadim* Talanov[1], and *Eremey* Valetov[1]

[1]Paul Scherrer Institut, Forschungsstrasse 111, 5232 Villigen PSI, Switzerland
[2]Institute for Particle Physics and Astrophysics - ETH Zürich, Otto-Stern-Weg 5, 8093 Zürich, Switzerland
[3]INFN Sezione di Pisa, Largo B. Pontecorvo 3, 56127, Pisa, Italy
[4]Dipartimento di Fisica dellUniversitá di Pisa, Largo B. Pontecorvo 3, 56127 Pisa, Italy

Abstract. Currently, PSI delivers the most intense continuous muon beam in the world with up to a few 10^8 μ^+/s. The High-Intensity Muon Beams (HIMB) project is developing a new target station and muon beamlines able to deliver 10^{10} μ^+/s, with a huge impact for low-energy, high-precision muon experiments. While the next generation of proton drivers with beam powers in excess of the currently achieved 1.4 MW still require significant research and development, the focus of HIMB is to improve the surface muon yield with a new target geometry and to increase capture and transmission with a solenoid-based beamline in order to reach a total efficiency of approximately 10 %. We present the current status of the HIMB project.

1 Introduction

High-intensity proton-driven meson factories have contributed to stringent tests of the current Standard Model (SM) of particle physics with a number of precision experiments.Within this framework, high-intensity muon beams have been employed as sensitive probes of Beyond SM (BSM) physics by searching for SM highly suppressed muon decays, such as $\mu^+ \to e^+\gamma$, $\mu^+ \to e^+(e^+e^-)$ and $\mu^-N \to e^-N$. These so-called golden channels in the charged Lepton Flavor Violation (cLFV) sector, if detected, would be a direct signal of BSM physics. Anti-muons have also found applications outside of particle physics such as aiding solid-state physics, chemistry and material science by using the muon spin spectroscopy method (μSR) to probe magnetic properties of matter down to nanometer scales.

At PSI, the High Intensity Proton Accelerator (HIPA) facility delivers one of the most powerful proton beams to target stations, 1.4 MW, to produce the most intense continuous muon beams in the world, with rates up to few 10^8 μ^+/s, granting PSI the lead at the intensity frontier in both particle physics [1] and condensed matter research with muons [2–7]. Such high rates are possible by exploiting 'surface muons', which are anti-muons generated by π^+ decay at rest inside the production target within a very narrow momentum range. This is due

*e-mail: giovanni.dal-maso@psi.ch

 https://doi.org/10.1051/epjconf/202328201012

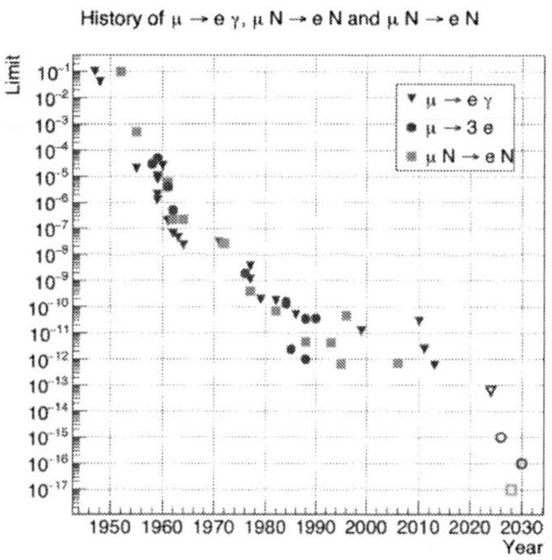

Figure 1. The history of limits for the cLFV golden channels branching ratios. The open markers indicate projected sensitivities. Plot modified from [8].

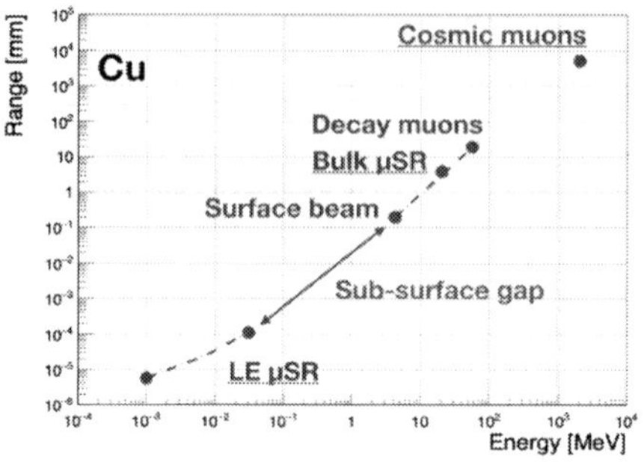

Figure 2. Calculated muon stopping ranges in Copper (Cu) as a function of the incoming muon energy. Plot modified from [9]. Typically used muon energies for μSR measurements restrict the probed ranges to either < 200 nm (low-energy muons, LE-μ^+) or ≥ 100 μm (surface- and decay-muon beams).

to the fact that they are created in a two-body decay, where the energy of the outcoming muon is only determined by the material budget it encounters inside the target. In addition, these muons are fully polarized.

The High-Intensity Muon Beams project (HIMB) at PSI aims at further pushing the current muon rates by two orders of magnitude up to 10^{10} μ^+/s, thus boosting reachable sensitivities for next-generation cLFV searches (Figure 1). Such high rates could be also exploited to improve limitations of the current μSR technique, such as the lack of intensity in the sub-surface region (see Figure 2), decreased measurement time and the possibility of probing multiple samples at once [9].

While next-generation proton drivers still require significant research and development to exceed the current limits, HIMB focuses on boosting surface muon yields with a new target station by increasing both capture and transport efficiency by building two new solenoid-based beamlines.

The HIMB project is part of the Isotope and Muon Production using Advanced Cyclotron and Target Technologies project [10] (IMPACT), whose other goal is the construction of a new radionuclide production target for an online isotope separation facility: the Targeted Alpha Tumour Therapy and Other Oncological Solutions (TATTOOS) project.

2 Target upgrade

Muons are produced at PSI by impinging the HIPA proton beam on two polycrystalline graphite rotating wheels: target M (TgM), which is 5.2 mm thick, and target E (TgE), which is 40 mm thick and devoted to high-intensity production. The rotation is used for radiative cooling and to distribute the heat load over the target rim.

To increase the surface muon yield, TgM station will be substituted with a high-intensity version, target H (TgH) with a slanted target geometry. Introducing a slanting angle while keeping the proton path length constant increases the surface muon yield as the pion production depends only on the material budget met by the protons in the target, while the surface-to-volume ratio of the target is increased as opposed to no slanting angle. TgH is going to be 20 mm thick in the proton beam direction, with a 10 deg slanting angle (see Figure 3):

 https://doi.org/10.1051/epjconf/202328201012

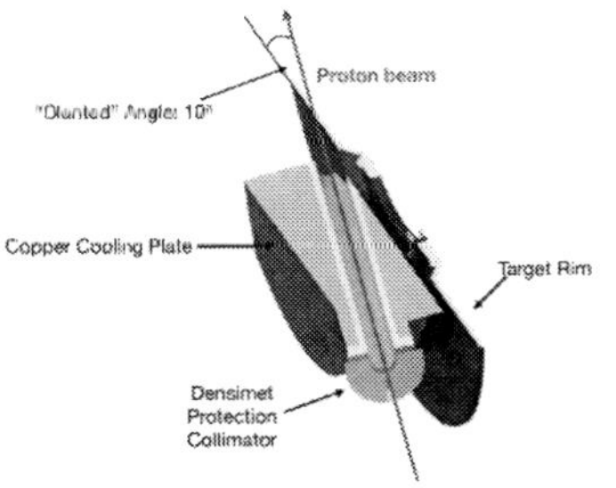

Figure 3. A section of the target TgH.

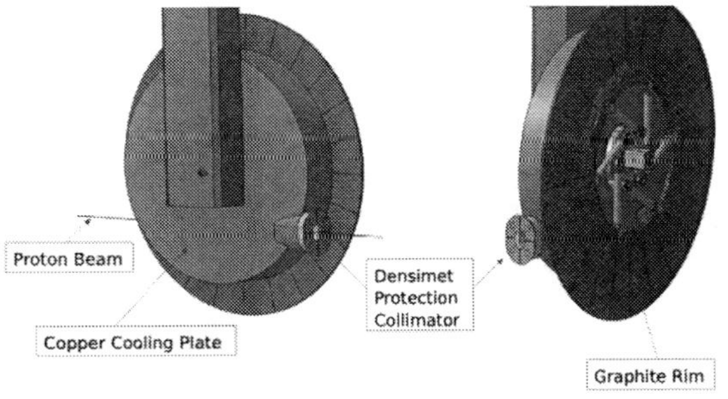

Figure 4. Full layout of the lower part of the TgH insert. The red line shows the trajectory of the proton beam, which impinges on the target from the back.

these parameters result in surface muon yields comparable to a 40 mm thick non-slanted target [11]. The slanted geometry has already been successfully tested at TgE, improving the surface muon yield by 40-50% depending on the beamline. The particles are planned to be extracted at 90 deg to the proton beam axis.

The proton beam will impinge on the back of the target, passing through a water-cooled copper plate acting as local shielding. A Densimet® (tungsten alloy) protection collimator will be used in order to protect the cooling plate.

To allow for such a configuration, the proton beam trajectory has to be lower than the rotation axis of the target, or it would intersect the rotation shaft. Additionally, a lower trajectory allows the use of a flat target, providing more usable space and further reducing the heat and radiation load on the target insert. Figure 4 shows a more detailed view of the TgH design.

The insert of TgH will fit into the existing TgE remote-controlled exchange flask (Figure 5). All elements that experience high radiation exposure along PSI beamlines are organised in inserts to allow for easier handling and maintenance. The extraction of those elements from the beamlines is performed using exchange flasks that provide shielding while moving the inserts in the experimental hall. The flasks are thick steel cases equipped with grippers and motors to insert and extract the different elements of the beamline. The possibility of handling both target stations with the same exchange flask allows for easier operations and reduces the number of needed spare components.

2.1 Capture

When using solenoids to capture particles produced at a target, the common approach is to have the target completely enclosed inside the solenoid aperture, as this provides maximum acceptance for the created particles. This solution is not straightforward to apply in the HIMB project as the proton beam is not going to be stopped at TgH: the proton beam must pass through both TgH and TgE to then be fully stopped in the spallation source of the Swiss Spallation Neutron Source SINQ. In order to avoid major changes to the proton beam optics to compensate for the high magnetic fields required for capture, our approach is to have two different capture solenoids sideways to the target. Figure 6 shows a schematic of the 'capture solenoid' together with the TgH.

Because of the high radiation exposure, the capture solenoids are chosen to be normal conducting. Mineral-insulated coils are used to withstand the high radiation load, and the water cooling is indirect to avoid the corrosion of the coils. The design is going to be similar to that already used in the μE4 beamline at PSI [12].

Figure 5. TgE remote-handling exchange flask lifted by the 60 t crane in the HIPA facility experimental hall.

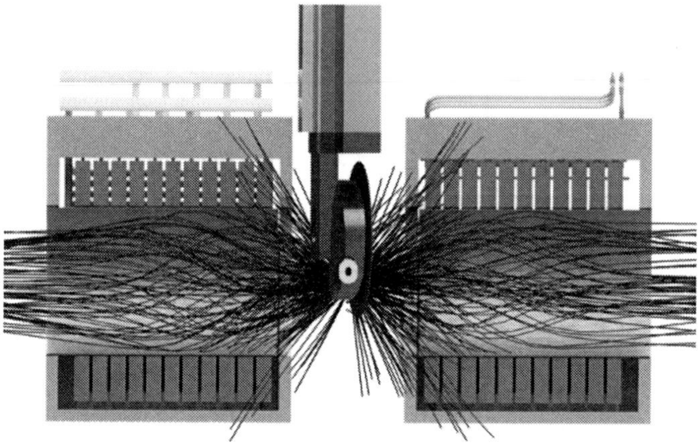

Figure 6. Schematic of the target region. The two capture solenoids are collecting muons (black lines) sideways to TgH.

The central field of the capture solenoids is planned to be $\sim 0.45\,\mathrm{T}$, and, to maximize capture, they are going to be positioned as close to the target as possible. Such a configuration will induce non-negligible fringe fields at TgH, and a dipole chicane will provide correction to the proton beam trajectory.

2.2 Particle production

The proton beam is accelerated to 590 MeV to impact TgH first and subsequently TgE. At such an energy, the only particles produced, together with the remnants of the nuclear reactions in the target, are electrons, muons and pions. Figure 7 shows the momentum spectrum of the relevant species produced at TgH.

As already introduced in Section 1, muons are produced through pion decay: the peak in the μ^+ spectrum in Figure 7 is caused by the surface muon production, while at higher energies, the muons come from pion decay-in-flight. In the case of μ^-, the surface muon feature is not present as there are no pion decays at rest: any stopped pion undergoes nuclear capture, and no muon is produced. Analogously, μ^+ stopped in the target decay at rest, producing a peak in the positron spectrum close to the Michel edge at $52.8\,\mathrm{MeV/c}$ [13].

The HIMB project focuses on increasing the delivered surface muon rates, but it will be possible as well to deliver the other particle species to the experimental areas. In order to do so, the dipoles will be designed to accept momenta up to $80\,\mathrm{MeV/c}$, but transmission at momenta higher than that of surface muons, $28\,\mathrm{MeV/c}$, will be limited by the maximum excitation currents of the solenoids.

3 Beamlines

Two new muon beamlines will be built within the HIMB project: MUH2 and MUH3. The main feature of these high-intensity muon beamlines is the extensive use of solenoids as focusing elements, as opposed to quadrupoles. The latter focus the beam in one transverse direction and defocus it in the other, solenoids provide focusing in both transverse directions simultaneously. This leads to higher capture capabilities, but usually, the required magnetic fields are much higher than for quadrupoles. For the momentum of the surface muons,

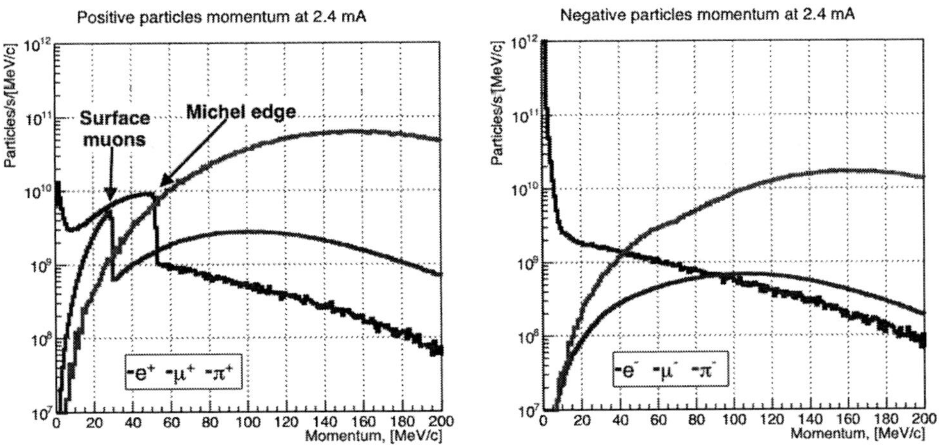

Figure 7. Momentum spectrum of positive (left) and negative (right) particles produced at TgH at a proton current of 2.4 mA [10].

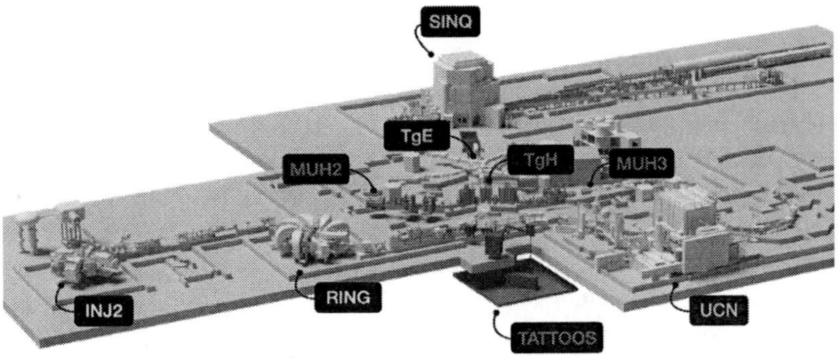

Figure 8. Overview of the HIPA accelerator complex including the HIMB beamlines and the new isotope production target station TATTOOS [10].

the required maximum fields for the capture and transport solenoids are ~ 0.45 T, which is achievable with normal conducting solenoids.

Figure 8 shows the layout of the HIPA experimental area, including the HIMB beamlines. The proton beam is bunched at 50 MHz and accelerated by the cyclotron Injector 2 (INJ2), then it is transported to the RING cyclotron for the final acceleration to reach 590 MeV energy. From there, the proton beam is transported to TgH. Additionally, the full beam can be kicked to the UCN facility or up to 100 μA can be scraped off to be delivered to TATTOOS. The HIMB beamlines will be oriented perpendicularly to the proton beamline. After TgH the proton beam impinges on TgE and is stopped at SINQ.

3.1 The MUH2 beamline

The MUH2 beamline is designed to deliver beams to fixed target experiments, e.g. Mu3e phase II [14]. It will be located at the left-hand side of the target station in the proton beam direction, with the beamline starting at a distance of 250 mm from TgH.

Being dedicated to particle physics, the most important figure of merit for MUH2 is transmission to be pushed as high as possible. To do so, the focusing elements along the full chan-

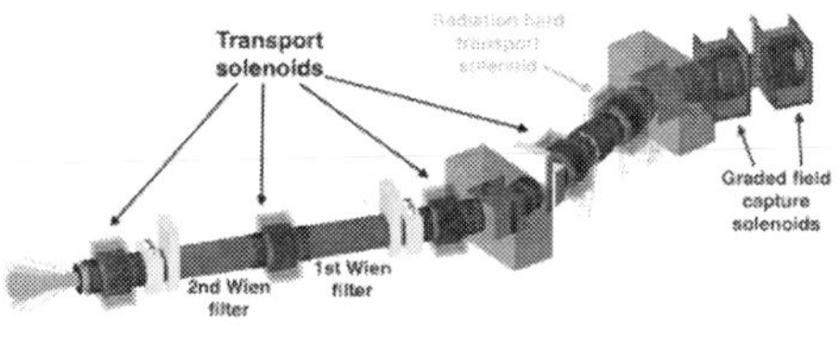

Figure 9. MUH2 beamline model in G4beamline [16] including the double Wien filter separation scheme and the graded-field capture solenoid. TgH is positioned at the intersection of the axes on the right. The muon beam (green lines) is captured from the capture solenoid and transported to the entrance of the experimental area located on the downstream side of the last transport solenoid on the left.

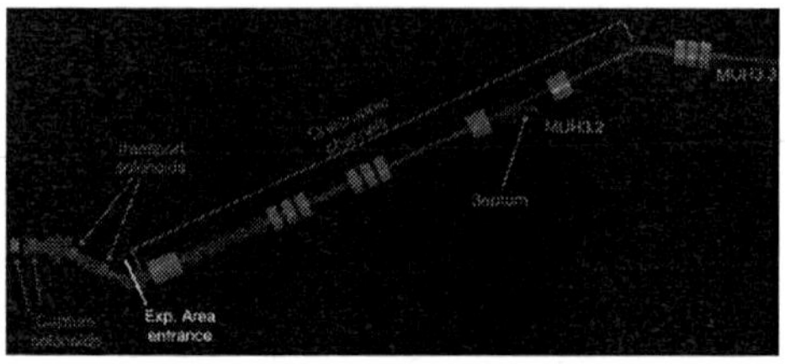

Figure 10. MUH3 beamline model in G4beamline [16]. TgH is positioned between the two capture solenoids on the left. The muon beam (blue lines) is delivered to the entrance of the experimental area by the solenoid-only section, to then continue through the quadrupole channels. The two branches MUH3.2 and MUH3.3 are served by a septum magnet.

nel will be solenoids. Another requirement for the layout of the beamline is to avoid a direct line of sight to the target due to the high radiation and neutron flux coming from the target: two 40 deg bends are included in the beam trajectory with two dipoles, leading to a stretched Z-layout. The initial optical model and design were studied using the matrix-based code Graphics Transport [15]. Then a G4beamline [16] model based on high-fidelity fieldmaps for each element was implemented.

The model reported in the Conceptual Design Report (CDR) published in January 2022 [10] is able to deliver 1.22×10^{10} μ^+/s at a proton current of 2.4 mA at the entrance of the experimental area at the surface muon momentum. The beam spot size and the average polarization at the end of the channel are $\sigma_x = 40$ mm, $\sigma_y = 42$ mm and 88 %.

As previously shown in Figure 7, at the target, the positron yield is roughly the same as that of anti-muons at 28 MeV/c, leading to non-negligible positron contamination at the entrance of the experimental area. Setting the MUH2 optics for surface muons leads to 1.94×10^{10} e^+/s at a proton current of 2.4 mA. The pion yield for the considered momentum instead is negligible. Figure 11 shows the delivered rates at the entrance of the experimental area at different momenta.

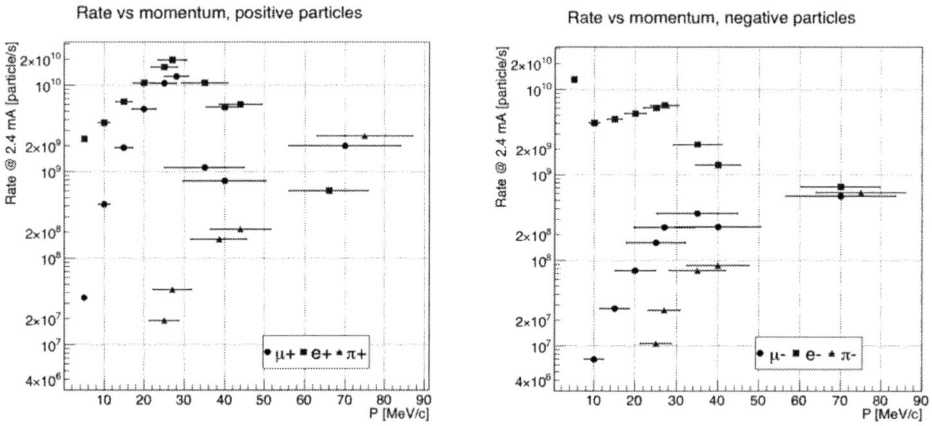

Figure 11. Momentum spectrum of positive (top) and negative (bottom) particles as delivered by MUH2 at the entrance of the experimental area for different momentum settings at a proton current of 2.4 mA [10]. The error bars show the standard deviation of the momentum distribution.

A double Wien filter scheme is currently under study to keep the positron contamination under control: the length of a single Wien filter to be effective in separating positrons and muons would not allow for an optimal transmission, leading to a muon rate reduction of 74 % [10]. By employing two Wien filters, separated by a solenoid to refocus the diverging beam, it is possible to reduce their lengths and increase the overall transmission. Detailed studies are still ongoing to find the optimal solution between contamination and transmission.

Another feature under study is the possibility of powering the coils of the capture solenoids independently to increase the absolute transmission of the beamlines. The current design of such a magnet, the 'graded-field solenoid', envisages three coils that can be excited independently to increase the degs of freedom at capture. Figure 9 shows the version of MUH2, including the double Wien filter scheme and the graded-field capture solenoid.

3.2 The MUH3 beamline

The MUH3 beamline aims at delivering muon beams for muon spin rotation spectroscopy (μSR). It will be located on the right-hand side of the target station in the proton beam direction, with a distance of 250 mm from TgH to the up-stream face of the coils in the capture solenoid. As introduced above, higher rates would be beneficial for such a technique to cover the sub-surface gap and increase the low energy muon (LEμ) rates. For such an application, the full 10^{10} μ^+/s rate is not required, as the needed statistics are lower than for particle physics experiments. Hence, the increase in rate with respect to the current μSR beamlines is provided by a first solenoid-only section up to the second bend, analogously to MUH2, while transmission to the experiments is given by a conventional quadrupole tuplet scheme. The MUH3 beamline will serve two experimental areas MUH3.2 and MUH3.3, that will be exploited alternatively or concurrently by employing a septum magnet. The studies on MUH3 optics are performed on Graphics TRANSPORT [15] and COSY INFINITY [17], while transport simulations are performed with a G4beamline [16] model including high-fidelity fieldmaps for each element. Figure 10 shows the G4beamline model of MUH3, including both branches.

The solenoid section delivers more than 10^{10} μ^+/s to the entrance of the experimental area at a proton current of 2.4 mA. Both MUH3.2 and MUH3.3 channels can deliver 3×10^8 μ^+/s to the experiments at the surface muon momentum, 3×10^7 μ^+/s at 15 MeV/c and 6×10^6 μ^+/s at 10 MeV/c, providing high enough rates for μSR measurements in the sub-surface gap. The polarizations in the considered cases are 97 %, 96 % and 95 %, respectively - comparable to that of the beams currently available at PSI [18][19]. The final focus optics are still under study, but we expect the final beam spots to be $\sigma_x \sim 20$ mm and $\sigma_y \sim 20$ mm.

The main focus for MUH3 will be on the surface and sub-surface muons. We are studying the beam properties at different momenta, such as rates and polarization.

4 Conclusions

The HIMB project is currently in a planning phase, with one out of three approval stages passed. Figure 12 shows the main milestones of our program:

- end of the preparation phase and publication of the Technical Design Report - end 2024
- a 1.5-year long shut-down - start 2027
- pilot operation - mid 2028
- novel user operation - mid 2029

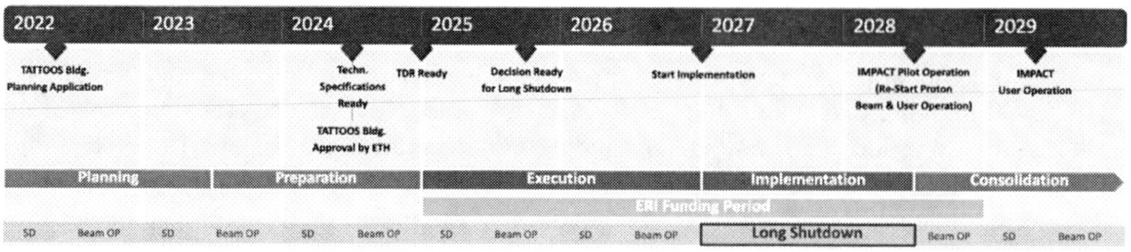

Figure 12. Timeline of the HIMB project including the main milestones.

The current design of the beamlines meets the primary goal of exceeding 10^{10} μ^+/s delivered rates, and we aim to prepare a final beamline layout with corresponding particle rates and envelopes by the end of 2022.

The authors thank the Paul Scherrer Institute as host laboratory and the technicians for their necessary support. This work was supported by the European Union's Horizon 2020 research and innovation programme under the Marie Skłodowska-Curie ITN Grant Agreement 858199 and COFUND Grant Agreement 884104.

References

[1] A. Signer et al., Particle Physics at PSI (SciPost Phys. Proc., 2021)
[2] A. Amato et al., Rev. Mod. Phys. **69-2**, 1119-1180 (1997)
[3] E. Morenzoni et al., Proceedings of the International Symposium on Science Explored by Ultra Slow Muon (1997)
[4] A. Amato et al., Physica C: Superconductivity **469-9**, 606-613 (2009)
[5] R. Khasanov et al., High Pressure Research **36-2**, 140-166 (2016)
[6] Z. Guguchia et al., Condensed Matter **5-2**, (2020)
[7] T. Shang, T. Shiroka, Frontiers in Physics **9**, (2021)
[8] R.H. Bernstein, Peter S. Cooper, Physics Reports **532-2**, 27-64 (2013)
[9] M. Aiba et al., Science Case for the new High-Intensity Muon Beams HIMB at PSI (2022)
[10] R. Eichler et al., PSI Bericht **22-01**, (2022)
[11] F. Berg et al., Phys. Rev. Accel. Beams **19-2**, 024701 (2016)
[12] T. Prokscha et al., Nucl. Instrum. Methods Phys. Res. A **595-2**, 317-331 (2008)
[13] J. Barlow et al., Proc. Phys. Soc. **84**, 239 (1964)
[14] A. Blondel et al., Research Proposal for an Experiment to Search for the Decay $\mu \to eee$ (2013)
[15] U. Rohrer, PSI Graphic Transport Framework by U. Rohrer, based on CERN-SLAC-FERMILAB version by K. L. Brown et al. (2006)
[16] Muons Inc., *G4beamline 3.06*, http://www.muonsinternal.com/muons3/G4beamline
[17] M. Berz, K. Makino, COSY INFINITY 9.0 Beam Physics Manual, Report number: MSUHEP-060804 (2006)
[18] PSI, *SμS: Swiss Muon Source*, https://www.psi.ch/en/smus
[19] A.M. Baldini et al., Eur. Phys. J. C **76**, 223 (2016)

https://doi.org/10.1051/epjconf/202328201013

Systematic effects in the search for the muon electric dipole moment using the frozen-spin technique

Chavdar Dutsov[1,*], *Timothy* Hume[1], and *Philipp* Schmidt-Wellenburg[1] on behalf of the muEDM collaboration

[1]Paul Scherrer Institute, Villigen 5232, Switzerland

Abstract. At the Paul Scherrer Institute (PSI) we are developing a high precision instrument to measure the muon electric dipole moment (EDM). The experiment is based on the frozen-spin method in which the spin precession induced by the anomalous magnetic moment is suppressed, thus increasing the signal-to-noise ratio for EDM signals to achieve a sensitivity otherwise unattainable using conventional $g-2$ muon storage rings. The expected statistical sensitivity for the EDM after a year of data taking is $6\times10^{-23}e{\cdot}$cm with the p = 125 MeV/c muon beam available at the PSI. Reaching this goal necessitates a comprehensive analysis on spurious effects that mimic the EDM signal. This work discusses a quantitative approach to study systematic effects for the frozen-spin method when searching for the muon EDM. Equations for the motion of the muon spin in the electromagnetic fields of the experimental system are analytically derived and validated by simulation.

1 Introduction

The presence of a permanent EDM in an elementary particle implies Charge-Parity (CP) symmetry violation. Even though the phase of the CKM matrix of the Standard Model of particle physics (SM) provides a large CP violating phase it results in tiny electric dipole moments of elementary particles, too small to measure any time soon. However, many SM extensions permit large CP violating phases, which also result in large electric dipole moments [1, 2]. Recently, the muon electric dipole moment has become a topic of particular interest due to the tensions between the measured muon anomalous magnetic moment and the SM expectation [3] and hints of lepton-flavor universality violation in B-meson decays [4, 5].

The frozen-spin technique, proposed by Farley et al. [6], is a method of measuring EDMs in storage rings where a radial electric field is applied to the stored particles such that the anomalous $(g-2)$ precession is cancelled, and any residual precession is due to the EDM. However, in a real storage ring the anomalous precession cannot be perfectly negated and an EDM-like precessions can be induced by coupling of the magnetic dipole moment (MDM) to the electromagnetic (EM) fields of the experimental setup – an example of a systematic effect.

We define systematic effects as all phenomena that lead to a real or apparent precession of the spin that are not related to the EDM.

2 Experimental setup

The search for the muon EDM will rely on a centimeter-scale storage ring inside a compact solenoid [7]. The muons will be injected into the solenoid one-by-one through a superconducting injection channel [8] and will be kicked by a pulsed magnetic field into a stable orbit

*e-mail: chavdar.dutsov@psi.ch

within a weakly focusing field [9]. The muon orbit will be positioned between two concentric cylindrical electrodes that will provide the radial electric field to deploy the frozen-spin technique. The direction of the muon spin at the time of its decay will be deduced from the direction of the emitted decay positron. This is made possible by their preferential emission, for high positron energies, in the direction of the muon spin due to parity violation in the weak decay. The EDM will be inferred from the change in asymmetry, $A = (N_u - N_d)/(N_u + N_d)$, between the number of positrons emitted upstream and downstream with respect to the direction of the muon beam. Upstream emission is determined by positrons emitted in the direction of the injection channels, while downstream emission is defined as those emitted in the opposite direction. The direction of emission will be monitored by detectors positioned symmetrically on both sides of the plane defined by the ideal muon orbit.

The collaboration is proceeding in a staged approach. In the initial phase we will demonstrate the feasibility of all critical techniques and aim for sensitivity better than 3×10^{-21} e·cm. In the final phase we target a sensitivity of better than $6 \times 10^{-23} e$·cm – an improvement of more than three orders of magnitude over the current experimental limit $1.8 \times 10^{-19} e$·cm (CL 95%) [10].

3 Limits on systematic effects

In order to evaluate systematic effects related to the EM fields in the experiment it is necessary to study the relativistic spin motion in electric $\vec{E}$ and magnetic $\vec{B}$ fields described by the Thomas-Bargmann-Michel-Telegdi (T-BMT) equation [11, 12]:

$$\vec{\Omega} = -\frac{e}{m_0}\left[a\vec{B} - a\frac{\gamma-1}{\gamma}\frac{(\vec{\beta}\cdot\vec{B})\vec{\beta}}{|\vec{\beta}|^2} + \left(\frac{1}{\gamma^2-1} - a\right)\frac{\vec{\beta}\times\vec{E}}{c} + \frac{\eta}{2}\left(\frac{\vec{E}}{c} + \vec{\beta}\times\vec{B}\right)\right]. \quad (1)$$

The coordinate system used throughout this work is such that it follows the reference particle orbit (similar to [13, 14]) and is sketched in Figure 1.

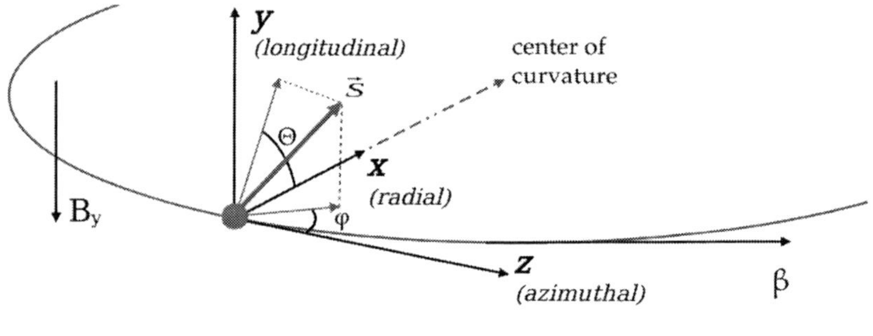

Figure 1. The local reference coordinate system used to derive the motion of the spin in the EM fields of the experiment. The vector $\hat{z}$ follows the momentum of the muon and $\hat{y}$ always points along the longitudinal direction defined by the main solenoid magnetic field B_y.

Using equation (1) and assuming that $\vec{\beta}\cdot\vec{E} = 0$ and $\vec{\beta}\cdot\vec{B} = 0$ the angle of spin precession around the radial axis due to a non-zero EDM is:

$$\Theta(t) = \Omega_{\hat{x}}^{\mathrm{EDM}} t = \frac{\eta}{2}\frac{e}{m_0}\beta_z B_y t, \quad (2)$$

where t is the elapsed time between the injection of the muon into the longitudinal magnetic field and its decay.

The relationship between the dimensionless parameter η that characterizes the spin precession and the EDM d is given by:

$$d = \frac{e\hbar}{4m_0c}\eta. \tag{3}$$

Combining (2) and (3) gives an expression of the rate of change of the angle of rotation of the spin $\dot{\Theta}$ as a function of the EDM d:

$$\dot{\Theta} = \frac{2c}{\hbar}\beta_z B_y d. \tag{4}$$

For the precursor and final experiments $\beta_z = 0.26$ and 0.77 respectively and $B_y = 3$ T, hence, the angular velocities are

$$\dot{\Theta} = 21.15 \ \mu\text{rad}/\mu\text{s}, \ \text{for}\ d = 3\times10^{-21}\ e\cdot\text{cm and} \tag{5}$$

$$\dot{\Theta} = 1.26 \ \mu\text{rad}/\mu\text{s}, \ \text{for}\ d = 6\times10^{-23}\ e\cdot\text{cm}. \tag{6}$$

In the following sections we will discuss various effects with the goal of deriving the full description of the spin's motion in the EM fields of the system. The derived equations can be used to place limits on measurement parameters such that the angular velocity of the spin precession due to the MDM is less than the experimental sensitivities given in (5) and (6).

4 Sources of real spin precession

One of the large sources of a false EDM signal are oscillations and rotations of the spin due to the coupling of the MDM of the muon with the electric and magnetic fields present in the experimental setup. As a starting point we will describe analytically the motion of the spin due to the combined effect of ideal approximations of the magnetic field of the solenoid, the weakly focusing field and the electric field used in the frozen-spin method. The magnetic field of the solenoid is approximated as a uniform magnetic field oriented along the $\hat{y}$ axis. The weakly focusing field is described by the approximated field that is generated by a circular coil. The electric field is assumed to be a radial field generated from the potential difference between two infinite coaxial cylindrical electrodes. Some possible and important imperfections of these fields and their effect on the spin precession is discussed.

4.1 Spin precession along the radial axis

Assuming the particles follow a trajectory with a constant radius ρ, the radial magnetic field $B_x(y)$ of a cylindrical coil with N turns and radius R along that trajectory can be approximated with [15]:

$$B_x(y) = \frac{3}{2}\mu_0 NIR^2\frac{\rho y}{(R^2+y^2)^{\frac{5}{2}}}, \tag{7}$$

where I is the current passing through the coil and μ_0 is the magnetic permittivity of vacuum. Expanding (7) in a Taylor series one obtains:

$$B_x(y) = \Phi_0\rho y - \frac{5}{2}\frac{\Phi_0}{R^2}\rho y^3 + O(y^5), \ \text{where}\ \Phi_0 = \frac{3}{2}\frac{\mu_0 NI}{R^3}. \tag{8}$$

The longitudinal position of a particle with charge e, mass m_0 and velocity $c\vec{\beta}$ is given by the solution of:

$$\ddot{y} = \frac{1}{\gamma m_0}\left(eE_y + c\beta_z B_x(y)\right), \tag{9}$$

where we assume a constant non-zero longitudinal component of the electric field. The solution of the differential equation is the harmonic oscillator:

$$y(t) = y_0 \cos(\omega_\beta t + \varphi) + \frac{1}{\Phi_0 \rho} \frac{E_y}{c\beta_z}, \tag{10}$$

$$\text{where } \omega_\beta = \sqrt{\Phi_0 \frac{ec\beta_z}{\gamma m_0} \rho} \tag{11}$$

is the angular velocity of the vertical betatron oscillations (VBO).

For the precursor experiment where $\beta = 0.258$ and $\gamma = 1.03$ the period $T = 2\pi/\omega_\beta$ of the VBO is approximately 600 ns. Note that this period depends only on the strength of the focusing field and the radial position of the muon. It does not depend on the momentum of the muon.

The precession of the spin due to the coupling of the MDM with the radial magnetic field due to the weakly focusing field is:

$$\left(\vec{\Omega}^{\text{WF}}_{\text{MDM}}\right)_x = -\frac{ea}{m_0} B_x(y(t)) == -\frac{ea}{m_0} \left[\Phi_0 \cos(\omega_\beta t + \phi_0) \rho y_0 - \frac{1}{c\beta_z} E_y \right]. \tag{12}$$

The other source of radial precession that has to be considered is the radial magnetic field in the reference frame of the muon due to the non-zero longitudinal electric field in the laboratory reference frame. Using the T-BMT equation and taking only the radial component of the spin precession due to the electric field one obtains:

$$\left(\vec{\Omega}^{E_y}_{\text{MDM}}\right)_x = -\frac{e}{m_0 c} \left(a - \frac{1}{\gamma^2 - 1} \right) \beta_z E_y. \tag{13}$$

Combining (12) and (13), for the total angular velocity of the radial precession due to the MDM around the $\hat{x}$ axis, one obtains :

$$\left(\vec{\Omega}_{\text{MDM}}\right)_x = -\frac{ea}{m_0} \left[\frac{1}{c} \left(1 - \frac{1}{a(\gamma^2 - 1)} - \frac{1}{\beta_z^2} \right) \beta_z E_y + \Phi_0 \cos(\omega_\beta t + \phi_0) \rho y_0 \right]. \tag{14}$$

4.2 Azimuthal spin precession

When the muons circulate in the storage ring they oscillate around an equilibrium orbit. Because of this oscillation the momentum of the particle is not at all times perpendicular to the longitudinal magnetic field leading to a non-zero projection of the magnetic field along its trajectory. This field is proportional to the angle between the muon momentum and the magnetic field $\delta = \angle(\vec{\beta}, \vec{B})$. In turn, $\cos(\delta) = p_y/p_z$ and p_y oscillates as $p_y = p_{y_0} \sin(\omega_\beta t)$. Thus:

$$B_z(t) = B_y \cos(\delta) \approx \frac{p_{y_0}}{p_z} \sin(\omega_\beta t) B_y. \tag{15}$$

The momentum p_{y_0} is the momentum in the $\hat{y}$ direction when the muon is on the equilibrium orbit. It can be calculated as:

$$p_{y_0} = ec\beta\Phi_0 \rho y_0 \int_0^{\frac{\pi}{2\omega_\beta}} \cos(\omega_\beta t) dt = ec\beta\Phi_0 \rho y_0 \frac{1}{\omega_\beta}. \tag{16}$$

In the ideal case $\vec{\beta} = |\beta|\hat{z}$ and $\vec{E} = |E|\hat{x}$, but as the muons oscillate in the weakly focusing field there will be a non-zero $\hat{y}$ component of the velocity, thus:

$$\beta_y = (p_{y_0}/p_z)\beta_z \sin(\omega_\beta t). \tag{17}$$

If the radial electric field E_x is correctly set to the value E_f required by the frozen-spin technique, then there will be no oscillations around the $\hat{z}$ axis, excluding the second order $\beta \cdot \vec{B}$ term in (1). At this value for the freeze field, it will counteract the precession induced by coupling of the MDM to the longitudinal field of the solenoid. However, if $E_x \neq E_f$ there will be an imperfect cancellation of the $g-2$ precession around $\hat{z}$ which is proportional to the excess electric field $E_{ex} = E^{(\text{real})} - E^{(\text{freeze})}$ that the muon is subjected to:

$$\left(\vec{\Omega}^{E_{ex}}_{\text{MDM}}\right)_z = -\frac{e}{m_0 c}\left(a - \frac{1}{\gamma^2 - 1}\right)\beta_y E_{\text{ex}}. \tag{18}$$

Considering (15), the angular velocity of the spin precession along the $\hat{z}$ axis due to the $\vec{\beta} \cdot \vec{B}$ term in the T-BMT equation is:

$$\left(\vec{\Omega}^{\beta \cdot B}_{\text{MDM}}\right)_z = \frac{ea}{m_0}\left(\frac{\gamma - 1}{\gamma}\right) B_z(t). \tag{19}$$

Combining equations (19) and (18) gives the total angular velocity of the precession due to the MDM around the $\hat{z}$ axis:

$$\left(\vec{\Omega}_{\text{MDM}}\right)_z = -\frac{e}{m_0}\frac{p_{y0}}{p_z} \sin\left(\omega_\beta t\right) \times \left[\left(a - \frac{1}{\gamma^2 - 1}\right)\frac{\beta_z}{c} E_{ex} - a\left(\frac{\gamma - 1}{\gamma}\right) B_y\right]. \tag{20}$$

4.3 Electric field imperfections

The value of $E^{(\text{real})}$ may not be constant throughout the orbit of the muon if the central axes of the magnetic and electric fields are displaced or inclined with respect to each other. To obtain the components of the electric field with respect to the central axis of the solenoid, first consider a purely radial electric field created by perfect coaxial cylindrical electrodes with radii A and B. In the reference frame (x', y', z') of the electrodes, where z' is parallel to the central axis of the electrodes, the field is given by:

$$\vec{E}'(\vec{r}) = \frac{V}{\log \frac{B}{A}} \begin{pmatrix} x'/r^2 \\ y'/r^2 \\ 0 \end{pmatrix}, \tag{21}$$

where $r^2 = (x')^2 + (y')^2$. The electrode central axis, however, may be displaced and rotated with respect to the reference frame defined by the solenoid central axis. The electric field $\vec{E}$ in the reference frame of the solenoid can be obtained by transforming $\vec{E}'$ as:

$$\vec{E} = R_y(\alpha)\vec{E}'(R_y^{-1}(\alpha)\vec{r} + \vec{r}_0), \tag{22}$$

where $\vec{r}_0 = (x'_0, y'_0, 0)$ is the displacement between the two fields. $R_y(\alpha)$ is the rotation around the y' axis at an angle α, where α is the angle between the central axis of the cylindrical electrodes and the central axis of the longitudinal magnetic field. Note here that due to the rotational symmetry of the fields we can always choose the reference frame to be such that arbitrary displacements can be represented in this way. The electric field in the reference frame defined by the longitudinal magnetic field is then:

$$\vec{E}(\vec{r}) = V_0 \begin{pmatrix} \frac{\xi}{\rho^2}\cos\alpha \\ \frac{\upsilon}{\rho^2} \\ -\frac{\xi}{\rho^2}\sin\alpha \end{pmatrix} \tag{23}$$

where $V_0 = V/\log(B/A)$, $\upsilon = y' + y'_0$, $\xi = x'_0 + x'\cos\alpha - z'\sin\alpha$ and $\rho^2 = \xi^2 + \upsilon^2$.

The average of the radial electric field over the circular orbit of the muon can be obtained if E' is represented in cylindrical coordinates (ρ, ϕ, ζ) as:

$$\langle E(\rho,\zeta)\rangle = \frac{1}{2\pi}\int_0^{2\pi} E d\phi, \tag{24}$$

where ρ is the radius of the orbit of the muon as a function of the magnetic field and the momentum of the particle, and ζ is parallel to $\hat{y}$.

The radial electric field the muon experiences in its reference frame is modeled as:

$$E^{(\text{real})} = \langle E(\rho,\zeta)\rangle_\rho + \frac{1}{2}\left(E_{\rho,max} - E_{\rho,min}\right)\cos(\omega_c t + \beta_0), \tag{25}$$

where $\omega_c = -eB_y/\gamma m_0$ is the cyclotron angular velocity, E_{max} and E_{min} are the maximal and minimal value of the electric field over one rotation of the muon and β_0 is the initial phase of the muon position along the orbit.

Note, (24) is valid only in the case of a circular orbit. In this case it can be shown numerically that:

$$\langle E(\rho,\zeta)\rangle = \langle E'(\rho,\zeta)\rangle, \tag{26}$$

i.e., the rotation of the anode or cathode with respect to the muon orbit does not influence the average frozen spin condition and, more importantly, does not change the net E-field component in the $\hat{y}$ direction. As the centripetal force due to the B-field is $\approx 10^3$ larger than that due to the E-field and the expected misalignments between the center of the orbit and the center of the inner electrode are small, the circular orbit approximation holds well.

4.4 Combined spin precession

The initial orientation of the spin $\vec{S} = (S_x, S_y, S_z)$ in spherical coordinates is:

$$\phi_0 = \arctan\left(\frac{S_x}{S_z}\right), \quad \Theta_0 = \arctan\left(\frac{\sqrt{S_x^2 + S_z^2}}{S_y}\right) - \frac{\pi}{2}. \tag{27}$$

If there is an imperfect cancellation of the $g-2$ precession then there will be a rotation of the spin around the $\hat{y}$ axis. Thus, the angular velocity of the spin precession around $\hat{x}$ and $\hat{z}$ will be a projection of (14) and (20) along $\hat{y}$:

$$|\vec{\Omega}_{\text{MDM}} \cdot \hat{y}| = \left(\vec{\Omega}_{\text{MDM}}\right)_x \cos(\omega_y t + \phi_0) + \left(\vec{\Omega}_{\text{MDM}}\right)_z \sin(\omega_y t + \phi_0), \tag{28}$$

where ω_y is the angular velocity of the precession around $\hat{y}$ due to the $g-2$ precession and is:

$$\omega_y = \frac{ea}{m_0}\frac{E_{\text{ex}}}{E_f}B_y. \tag{29}$$

In practice, E_{ex} will oscillate (see (25)) with the cyclotron frequency due to the changing distance between the muon and the E-field center. The longitudinal B-field B_y will oscillate with the VBO frequency due to the changing longitudinal component of the weakly focusing field (proportional to (7)). In a well tuned frozen-spin experiment ω_y is much lower than ω_β and ω_c, and in this derivation the rotation of the spin around $\hat{y}$ is approximated with a constant angular velocity using the average values of E_{ex} and B_y.

The total longitudinal rotation of the spin Θ is:

$$\Theta(t) = \int_0^t |\Omega_{\text{MDM}} \cdot \hat{y}| dt \tag{30}$$

Calculating (30) for the $\hat{z}$ component only one obtains:

$$\Theta_{\hat{z}}(t) = -\frac{e}{m_0} \int_0^t \left[\frac{p_{y_0}}{p_z} \left(\frac{\beta_z}{c} \left(a - \frac{1}{\gamma^2 - 1} \right) E_{\text{ex}} - a \left(\frac{\gamma - 1}{\gamma} \right) B_y \right) \sin(\omega_\beta t') + aB_z \right] \sin(\omega_y t' + \phi_0) dt' \tag{31}$$

Noting that:

$$\int_0^t \sin(\omega_\beta t') \sin(\omega_y t' + \phi_0) dt' \approx -\frac{1}{\omega_\beta} \cos(\omega_\beta t) \sin(\omega_y t + \phi_0), \quad \text{if } \omega_\beta >> \omega_y, \tag{32}$$

the oscillations due to the excess electric field along the $\hat{z}$ axis (along the momentum) and due to the non-zero B_z magnetic field can be calculated by:

$$\Theta_{\hat{z}}(t) = \frac{e}{m_0} \left[\frac{p_{y_0}}{p_z} \frac{1}{\omega_\beta} \cos\left(\omega_\beta t\right) \left(\frac{\beta_z}{c} \left(a - \frac{1}{\gamma^2 - 1} \right) E_{\text{ex}} - a \left(\frac{\gamma - 1}{\gamma} \right) B_y \right) - \frac{1}{\omega_y} aB_z \right] \sin(\omega_y t + \phi_0). \tag{33}$$

A similar calculation can be performed for the spin-precession along the $\hat{x}$ axis due to the radial magnetic field of the weakly focusing field:

$$\Theta_{\hat{x}}(t) = -\frac{ea}{m_0} \left[\frac{1}{\omega_\beta} \Phi_0 \sin(\omega_\beta t + \phi_0) \rho y_0 \right] \cos(\omega_y t + \phi_0). \tag{34}$$

The spin-precession due to a non-zero electric field in the $\hat{y}$ direction can be easily integrated without approximations and gives:

$$\Theta_{\hat{z}\times\hat{y}}(t) = -\frac{ea}{m_0} \left[\left(1 - \frac{1}{a(\gamma^2 - 1)} - \frac{1}{\beta_z^2} \right) \frac{\beta_z}{c} E_y \right] \times \frac{1}{\omega_y} \sin(\omega_y t + \phi_0). \tag{35}$$

The total azimuthal angle Θ is the sum of the oscillations and rotations given above:

$$\Theta = \Theta_{\hat{x}} + \Theta_{\hat{z}} + \Theta_{\hat{z}\times\hat{y}} + \Theta_0. \tag{36}$$

Input parameters

The analytical equations have a few input parameters which can be roughly divided into two groups: stochastic initial conditions that vary for each particle in the storage ring at time t_0 of the experiment and constant or slowly changing parameters of the experimental system. The stochastic conditions in the first group are:

$$\begin{aligned} &\vec{p}_0 = (p_x, 0, p_z) - \text{Initial momentum of the particle.} \\ &\vec{S}_0 = (S_x, S_y, S_z) - \text{Initial spin of the particle.} \\ &\vec{r}_0 = (r_x, r_y, r_z) - \text{Initial position of the particle.} \\ &t_0 - \text{Time after injection for which } p_y = 0. \end{aligned} \tag{37}$$

Note here, that the p_y component is equal to zero as this is the defining moment when the particle is considered stored in the storage ring. The parameters r_x and r_z as well as p_x and p_z define the orbit of the muons and determine the electric field $E^{(\mathrm{real})}$ to which they are subjected. The parameter r_y is the starting position of the muon along the solenoid axis and defines the oscillations due to the radial weakly focusing magnetic field. The initial spin direction $\vec{S}_0$ determines the intermixing between the rotations around the $\hat{x}$ and $\hat{z}$ axes. There is also one constraint imposed on the spin orientation: $|S_0| = 1$.

The parameters of the experimental system are:

1. B_y – Main solenoid magnetic field strength.

2. I, R, N – Current through the N turns of the weakly focusing coils with radius R.

3. V, b, a, α – Voltage between the concentric cylinders with radii b and a defining the weakly focusing field. The angle α between the central axes of the electric field and solenoid magnetic field.

4. $p_0^{(\mathrm{ideal})}$ – Ideal momentum of the muons for which the freeze field is set.

5. E_y – Constant electric field normal to the storage ring plane.

Thus, the analytical equations depend on 8 free stochastic parameters and 10 constant parameters of the experimental system. These parameters completely define the dynamics of the spin inside this model of a storage ring.

4.5 Comparison with Geant4 spin tracking

In order to verify the analytical equations a model of the experiment was developed using the Geant4 Monte Carlo simulation toolkit [16]. The EM fields of the experiment can be either calculated analytically or interpolated from field maps. The field maps are generated by ANSYS Maxwell finite element modeling (FEM) software. The simulation has three major EM field components:

1. The main solenoid magnetic field – constant value along $\hat{y}$ or field map supplied by FEM simulations.

2. Radial electric field given by (23) with a possibility to add a constant and uniform component in the longitudinal direction or FEM simulated field map.

3. Weakly focusing field modeled in ANSYS as a single circular coil with $R = 65$ mm.

The muons start with zero momentum in the $\hat{y}$ direction as this is the initial condition for a stored muon. The simulation tracks the spin orientation in the reference frame of the muon and records it as a function of time. It can also track the direction with respect to a reference frame defined by the experimental setup, e.g., the solenoid.

A comparison between the derived analytical equations and the Geant4 spin tracking is shown in Figure 2. In the comparison a fine field map (0.2 mm step size) of the weakly focused field was used. All other EM fields are calculated analytically. The stepper used is the DormandPrinceRK78 routine [16] with 0.26 mm step size. The stepping size was chosen so as to avoid effects due to resonances between the stepper and field map grids. The initial coordinates of the muon at the moment of storage were chosen arbitrarily (values specified in Figure 2 caption) for the purpose of the illustration. The electric field was set to such a value as to have imperfect cancellation of the $(g-2)$ precession. The anode coaxial E-field

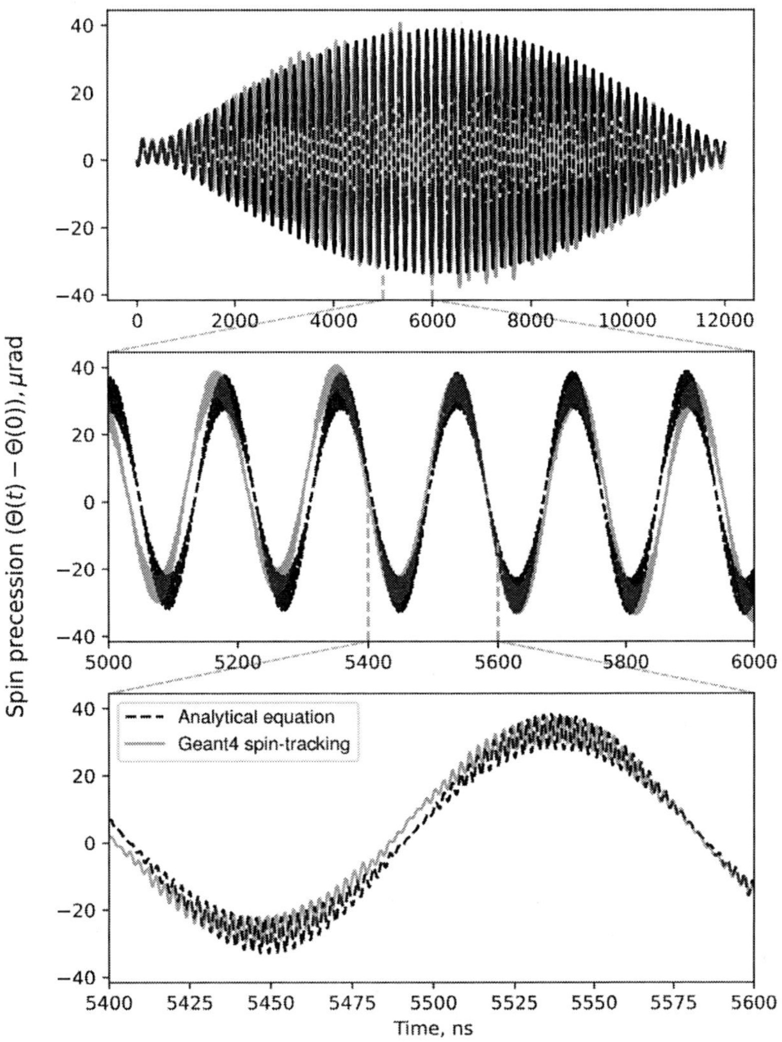

Figure 2. Comparison between the analytical equations (36) and the Geant4 spin tracking simulation. The initial parameters are arbitrarily set to $\vec{S}_0 = (-0.89, 0.00, -0.46)$, $\vec{p}_0 = (0.84, 0.00, 0.55) \times$ 26.8 MeV/c and $\vec{r}_0 = (-16.95, 5.00, 22.10)$ mm. The ideal momentum for which the freeze field is set is $p_0^{(\text{ideal})} = 28.0$ MeV/c. The three subfigures show different timescales of the spin motion – uncompensated $(g-2)$ precession, betatron oscillations and cyclotron oscillations.

is tilted with respect to the longitudinal axis at $\alpha = 0.01°$ in order to highlight the cyclotron oscillations.

The comparison shows very good agreement between the analytical equations and the Geant4 spin tracking. The approximate equation for the description of the weakly focusing field (8) provides good estimates of the field strength and VBO frequency. The difference between the analytical and numerical approaches is less than 5 μrad sustained over 12 μs of simulation time, demonstrating the very good agreement between the two also when using realistic FEM generated field maps in Geant4.

5 Sources of apparent spin precession

If the muon spin has a component in the longitudinal direction then the probability for up-stream p_u and downstream p_d ejected decay positrons will differ and an asymmetry A will be observed:

$$A(t) = \frac{p_u - p_d}{p_u + p_d} = \sin(\Theta(t))\,\alpha P \approx \Theta(t)\,\alpha P, \tag{38}$$

where α is the parity violating decay asymmetry averaged over the positron energy and P is the initial polarization. Equation (38) is valid when $\Theta(t) \ll 1$, which is the case for an EDM induced spin precession.

Using (2) and (38), the rate of change of the asymmetry $\dot{A}$ is:

$$\dot{A} = \frac{\eta}{2}\frac{e}{m_0}\beta_z B_y \alpha P. \tag{39}$$

We would require that the measured asymmetry $\dot{A}_m$ due to spurious EDM mimicking effects is $\dot{A}_m \ll \dot{A}$.

The number of detected positrons in the upstream detector N_u is given by:

$$N_u = \Omega_u \varepsilon_u p_u D, \tag{40}$$

where D is the total number of decayed muons, ε_u is the detection efficiency for positrons and Ω_u is the solid angle coverage of the upstream detector. Both Ω_u and ε_u can be summarized with a single parameter $\kappa_u = \Omega_u \varepsilon_u$ expressing the effective detection efficiency of the upstream detector. A similar equation can be given for the downstream detector and so:

$$N_u = \kappa_u p_u D = \kappa_u D_u \text{ and } N_d = \kappa_d p_d D. \tag{41}$$

From the point of view of the experiment, substituting equation (41) in (38), the measured asymmetry A_m is given by:

$$A_m = \frac{1}{D}\left(\frac{N_u}{\kappa_u} - \frac{N_d}{\kappa_d}\right). \tag{42}$$

If the effective detection efficiencies κ are constant in time no systematic effect appears, but a time dependence, e.g., by a disturbance of the detector electronics through the pulsed kicker field, leads to a systematic bias. A worst-case scenario assumption is the case where the kicker field disturbs the detection efficiency in both the upstream and downstream detectors in opposite directions and its effect reduces exponentially with time. The disturbance can be modeled as:

$$\kappa_u = \kappa_{u0} - \Delta_\kappa e^{-t/\tau_k} \text{ and } \kappa_d = \kappa_{d0} + \Delta_\kappa e^{-t/\tau_k}, \tag{43}$$

where κ_{u0} and κ_{d0} are some equilibrium detection efficiencies, Δ_κ is the perturbation in the efficiency due to the kicker field and τ_k is the time-constant of its effect.

In order to quantify the false EDM signal that can be caused by time-dependent efficiency parameters let us assume that there is no true EDM signal and the upstream and downstream emission probabilities are equal $p_u = p_d$ leading to $D_u = D_d = D/2$. Then the observed asymmetry is:

$$A_m = \kappa_u - \kappa_d = \kappa_{u0} - \Delta_\kappa e^{-t/\tau_k} - \kappa_{d0} - \Delta_\kappa e^{-t/\tau_k}. \tag{44}$$

The measured EDM signal is given by the time derivative of the asymmetry, thus:

$$\dot{A}_m = \frac{2}{\tau_k}\Delta_\kappa e^{-t/\tau_k}. \tag{45}$$

 https://doi.org/10.1051/epjconf/202328201013

From equation (45) we can observe that the systematic effect decreases with time t as expected. The effect is exacerbated if τ_k is on the order of magnitude of the experimental measurement time t which is several muon lifetimes. The limits on τ_k and Δ_κ can be derived by requesting that the measured false asymmetry A_m is lower than the theoretical prediction $\acute{A}$ from (39) for a given η.

6 Conclusions

Analytical equations for the description of the MDM precession in the EM fields of the experiment were derived. The equations were compared to a Geant4 Monte Carlo simulation using realistic field maps generated using the FEM software ANSYS Maxwell. Very good agreement has been observed between the two and the analytical equations seem to describe well the spin motion at short, medium and long timescales.

A method of calculation of the measured false EDM signal due to exponential changes in the positron detection efficiency is shown. If the detection efficiency is constant, even if different, for each detector for early-to-late times, then there would be no false EDM signal. A false EDM would be generated in the case of change in the detection efficiency of upstream relative to downstream detectors and the systematic effect is more prominent for time-constants on the order of the muon lifetime.

The presented methods will be used to derive limits on the mechanical production and EM field uniformity of the muon EDM experiments to be built at the PSI.

This work is financed by the Swiss National Science Fund under grant № 204118.

References

[1] M. Pospelov, A. Ritz, Ann. Phys. **318**, 119 (2005)
[2] C.Y. Seng, Phys. Rev. C **91** (2015)
[3] B. Abi, T. Albahri, S. Al-Kilani, D. Allspach, L. Alonzi, A. Anastasi, A. Anisenkov, F. Azfar, K. Badgley, S. Baeßler et al., Phys. Rev. Lett. **126** (2021)
[4] W. Altmannshofer, C. Niehoff, P. Stangl, D.M. Straub, Eur. Phys. J. A **77** (2017)
[5] B. Capdevila, A. Crivellin, S. Descotes-Genon, J. Matias, J. Virto, J. High Energy Phys. **2018** (2018)
[6] F.J.M. Farley, K. Jungmann, J.P. Miller, W.M. Morse, Y.F. Orlov, B.L. Roberts, Y.K. Semertzidis, A. Silenko, E.J. Stephenson, Phys. Rev. Lett. **93** (2004)
[7] A. Adelmann, M. Backhaus, C.C. Barajas, N. Berger, T. Bowcock, C. Calzolaio, G. Cavoto, R. Chislett, A. Crivellin, M. Daum et al., *Search for a muon EDM using the frozen-spin technique* (2021), `https://arxiv.org/abs/2102.08838`
[8] D. Barna, Phys. Rev. Accel. Beams **20** (2017)
[9] H. Iinuma, H. Nakayama, K. Oide, K. Sasaki, N. Saito, T. Mibe, M. Abe, Nucl. Instrum. Meth. A **832**, 51 (2016)
[10] G.W. Bennett, B. Bousquet, H.N. Brown, G. Bunce, R.M. Carey, P. Cushman, G.T. Danby, P.T. Debevec, M. Deile, H. Deng et al., Phys. Rev. D **80** (2009)
[11] L.H. Thomas, Lond. Edinb. Dublin Philos. Mag. **3**, 1 (1927)
[12] B. Montague, Phys. Rep. **113**, 1 (1984)
[13] M.H. Tahar, C. Carli, Phys. Rev. Accel. Beams **24** (2021)
[14] I.I. Rabi, N.F. Ramsey, J. Schwinger, Rev. Mod. Phys. **26**, 167 (1954)
[15] P.N. Murgatroyd, Am. J. Phys. **59**, 949 (1991)
[16] S. Agostinelli, J. Allison, K. Amako, J. Apostolakis, H. Araujo, P. Arce, M. Asai, D. Axen, S. Banerjee, G. Barrand et al., Nucl. Instrum. Meth. A **506**, 250 (2003)

EPJ Web of Conferences **282**, 01014 (2023) https://doi.org/10.1051/epjconf/202328201014
SSP 2022

Status of the COMET experiment

Andreas Jansen[1,*]
on behalf of the COMET collaboration

[1]TU Dresden, Institut für Kern- und Teilchenphysik, Germany

Abstract. The COMET (COherent Muon to Electron Transition) experiment, currently being built in Tokai, Japan, will search for the coherent neutrinoless transition of muons to electrons in the coulomb field of atomic nuclei. The process is highly suppressed in the Standard Model and therefore provides a promising channel to probe new physics. The experiment will be carried out in a staged approach. Phase-I of COMET aims to search for the process with a single event sensitivity of $\mathcal{O}(10^{-15})$. Additionally, precise measurements of muon beam dynamics and detector prototyping for Phase-II will be conducted. Utilizing a much higher intensity proton beam, a more complex and longer muon/electron transport system, and gained experience from Phase-I, an improvement of at least four orders of magnitude over the current best branching ratio limit up to $\mathcal{O}(10^{-17})$ is envisioned for Phase-II. This article will give a status report for Phase-I of COMET.

1 Introduction

The Standard Model (SM) of particle physics has already prevailed against various experimental tests for five decades – past and recent discoveries, like the SM-like Higgs boson, further crown its success. We know, however, from several observations that physics beyond the Standard Model (BSM) must exist. By way of example, the observed baryon-antibaryon asymmetry, a missing candidate for a dark matter particle, or the existence of neutrino mixing give a strong motivation for BSM physics. The latter example directly motivates the existence of small but non-zero neutrino masses and the violation of lepton flavor conservation.

One prominent charged lepton flavor violating (cLFV) process is $\mu - e$ conversion. This process could appear in the SM (including its extension with massive neutrinos); however, due to the tiny neutrino masses, it is strongly suppressed with a branching ratio of $< 10^{-54}$. Considering two widely acknowledged and well-motivated BSM theories, supersymmetry and warped extra dimensions, the cLFV effect is enhanced naturally up to around $< 10^{-14}$ [1, 2]. Consequently, discovering any cLVF process would be clear evidence for BSM physics and proves a powerful tool for BSM physics investigation.

The current best upper limit on $\mu{-}e$ conversion comes from the SINDRUM-II experiment, which set a limit on the conversion rate in muonic gold at 7×10^{-13} (90 % C.L.) [3]. The COMET experiment will search for the process with a 100 and 10000 times higher sensitivity; in its Phase-I and Phase-II, respectively. This article will focus on COMET Phase-I and aims to give an outline of the experiment, the current detector development status, and facility updates.

*e-mail: andreas.jansen@tu-dresden.de

2 $\mu - e$ conversion signal and backgrounds

To observe $\mu - e$ conversion, one starts by stopping negatively charged muons in material. Once the muon loses all its kinetic energy, it is captured by an atom into a high orbital momentum state. It will rapidly ($\sim 10^{-13}$ s) cascade down towards lower energy levels, ending in the 1s ground state of the muonic atom [4]. In the scope of BSM physics, the muon can now undergo the so-called neutrinoless coherent muon to electron conversion

$$\mu^- N(A, Z) \rightarrow e^- N(A, Z). \tag{1}$$

While the total lepton number L is conserved, individual lepton flavors L_e and L_μ are violated by one unit. The nucleus remains in its ground state and only takes up recoil in the conversion process - hence the reaction is called coherent. With the nucleus remaining unchanged, essentially only two particles are involved in the interaction. Therefore the energy of the outgoing electron has a fixed value E_e, given by the equation:

$$E_e = m_\mu - E_{Binding} - E_{Recoil}. \tag{2}$$

For lighter elements ($Z < 20$), the binding energy $E_{Binding}$, as well as the recoil energy of the nucleus E_{Recoil}, is small compared to the muon mass m_μ. The resulting energy of the electron, for example for an aluminum stopping target ($Z = 13$), is at $E_e = 104.97\,\text{MeV}$ [5]. Therefore, the conversion electron's energy is firmly positioned above most SM backgrounds. The remaining background contributions are small but must not be disregarded. For the sake of simplicity and to motivate experimental design decisions discussed in the next section, they are divided into three groups:

- **Cosmics:** One of the most critical backgrounds, cosmic rays, mimics the conversion signal by decay in flight or interaction with materials in the vicinity of the muon-stopping target. A cosmic ray veto system surrounding the whole detector region will be used to mitigate this background.
- **Intrinsic physics:** The highest intrinsic physics background stems from muon decay in orbit (DIO) events. While the electron momentum in a free muon decay is limited to roughly half the muon mass, the nuclear recoil in DIO allows the electron to surpass this limit significantly [5]. With the endpoint of the DIO spectrum reaching up to the conversion electron energy, only an excellent momentum resolution of the detector system can suppress this contribution.
- **Beam related:** The last group of backgrounds is caused by particles traveling along the muon beam line, such as electrons from muon and pion in-flight decays or antiprotons from the pion production target. With the half-life of light muonic atoms lying around multiple hundred ns, this prompt background can be strongly suppressed by implementing a delayed measurement time window. The resulting experimental requirements are a bunched beam structure, a high bunch purity, and a light material for the muon-stopping target.

3 The COMET experiment

An experimental overview of the COMET Phase-I layout is shown in Fig. 1. A pulsed 8 GeV proton beam impinges on a fixed target and produces charged pions which subsequently decay into muons (see left side of Fig. 1). They are guided along a short C-shaped beamline into the detector region. Here, two different configurations will be utilized. The configuration on the top right of Fig. 1 depicts the primary COMET Phase-I detector used to study the $\mu - e$ conversion process. The second configuration, shown on the bottom right, will be used to precisely measure beam dynamics, verify Monte-Carlo simulation, and deepen the understanding of the production system. A more detailed description and current status of the experimental parts are discussed in the following.

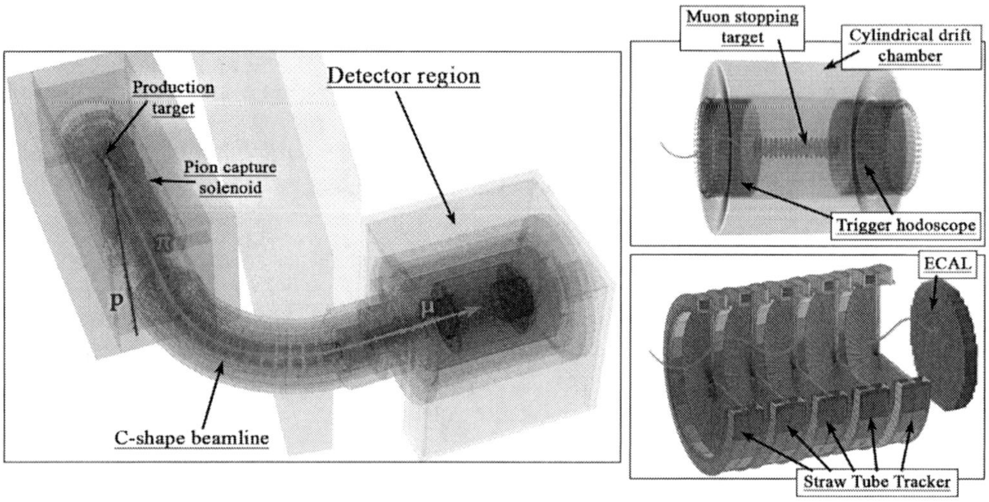

Figure 1. *Left:* Outline of the COMET Phase-I layout. For the detector region, two different configurations will be utilized. CyDet configuration (*top-right*); used to search for $\mu - e$ conversion. StrECAL configuration (*bottom-right*); used to measure the beam profile and prototype Phase-II detectors.

3.1 Proton beam and target

COMET is built at the Japan Proton Accelerator Research Complex (J-PARC), from where it will receive its 8 GeV proton beam. The beam energy is chosen to minimize production of antiprotons, which increases rapidly for proton energies above 10 GeV and could produce a fake 105 MeV electron signal. A major milestone achieved in 2022 is the completion of all beamlines leading from the J-PARC main ring to the COMET hall.

As motivated in section 2, COMET also requires a bunched beam to suppress beam-related backgrounds. Operation of the beam with an interval of 1.17 μs has successfully been performed. At the same time, the bunch purity was investigated by measurement of the so-called extinction factor, defined as *Number of residual protons between consecutive bunches* over *Number of protons in a bunch*. To achieve the experiment's design sensitivity, the extinction factor must be smaller than 10^{-10}. It was already measured at the J-PARC main ring position and shown to be less than 6×10^{-11}, which satisfies the requirements [6]. A second measurement at the COMET hall utilizing the finished beamlines will be performed in early 2023.

A mock-up of the graphite proton target for COMET Phase-I was produced and is shown in Fig. 2. Recent analysis hints at a larger-than-expected interference between the holding structure and the muon transport, but new design ideas are underway to mitigate this problem. The proton target will be installed inside the pion capture solenoid, designed to optimize the rate of low-energy negative pions produced in the backward direction. The outer hull is finished and also shown in Fig. 2. Construction of the whole setup is currently underway.

3.2 Cylindrical Detector system (CyDet)

The task of measuring the conversion signal in COMET Phase-I lies in the hands of the CyDet. It consists of a Cylindrical Drift Chamber (CDC), measuring the conversion signal electron, combined with a ring of scintillators used to provide the trigger and t_0 for each event. The whole system is contained in a 1 T solenoid field, and its inner radius of about 50 cm is

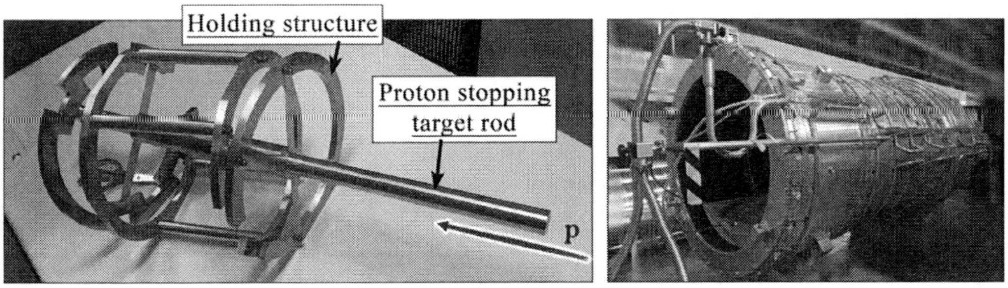

Figure 2. Status of parts in the pion production region. *Left:* Mock-up of the Phase-I proton target. *Right:* Outer hull of the pion capture solenoid.

designed to let most of the low momentum beam flash pass through the middle, avoiding a dangerously high hit rate in the detectors. Furthermore, a large part of the muonic atom DIO spectrum originating in the central aluminum muon-stopping target produces electrons with energies below 60 MeV which will not be able to reach the detector.

The CDC consists of 20 sense layers, each containing about 800 to 1200 wires. The CDC hull, subsequent wire stringing of all 19548 wires, tension measurements, etc., were finished in 2016, and pictures are shown in Fig. 3. Since then, various tests utilizing cosmic rays have taken place. They show an excellent performance of the CDC, with a measured spatial resolution of better than $\sim$ 200 μm. The result satisfies COMET design constraints, limiting the intrinsic physics background from high-energy DIO events to less than 0.01 events during the whole Phase-I runtime. The cosmic ray tests also demonstrate a successful stable operation of the whole chamber over more than two years.

Recently achieved milestones for the CDC are the complete installation and tests of the entire data acquisition system in 2019 and the design of a cooling system for the electronics. Performance tests for the latter were finished in early 2022, demonstrating that the temperature can be kept at sufficiently low 35 °C to limit FPGA heating.

Figure 3. Impression of the finished CDC setup. *Left:* CDC chamber after the stringing of all 19548 wires. *Right:* Picture of the drift chamber, including its fully installed data acquisition system.

The second part of CyDet is the CyDet Trigger Hodoscope (CTH). CTH rings are located at both ends of CDC, as shown at the top right of Fig. 1. A conversion electron will first spiral through the CDC, allowing its high precision momentum measurement, and subsequently hit the CTH, providing COMET's primary trigger. The CTH layout is depicted in Fig. 4. Each ring is segmented into 64 scintillation counters, with each counter tilted by a specific angle and shifted roughly half its width against the previous counter. A four-fold coincidence between scintillators will be utilized to reduce fake triggers caused by γ-rays coming mostly from Bremsstrahlung of DIO Michel electrons. With about 100 kHz, the CTH trigger rate is still expected to be very high and dominated by backgrounds. An online trigger selection, including CDC information, will be utilized to allow manageable data rates.

Development of the CTH is still ongoing. The right side of Fig. 4 shows a section of the full-sized prototype of the holding structure, which was recently produced to test the installation and fixation of scintillators. Currently, several studies using 1:1 scale counters are being carried out. Tests regarding timing, resolution, and photon yield were performed during summer 2022, using 100 MeV electrons at the Australian synchrotron. A detailed discussion of results is beyond the scope of this article, but overall the tests show good performance, satisfying COMET's requirements.

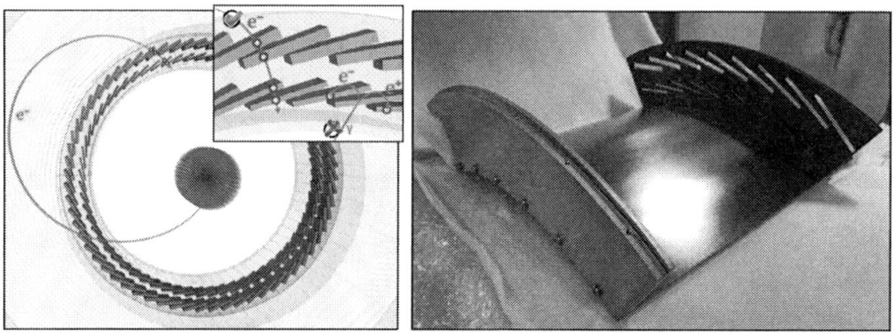

Figure 4. *Left:* Positioning of CTH counters inside CyDet. The inset demonstrates the required four-fold coincidence. *Right:* Impression of a section of the full-sized CTH prototype holding structure.

3.3 Straw-Tracker ECAL Detector System (StrECAL)

Instead of CyDet, a second configuration composed of a straw tracker and a crystal electromagnetic calorimeter, StrECAL, will be used after physics data taking. Its primary purpose is to directly measure the beam to enhance understanding of beam profile, particle production rate as a function of energy, and solenoid characteristics. The measurement will lay the foundation for the upcoming Phase-II extension of the transport solenoid. Moreover, at the end of this extension, a similar StrECAL system will replace the CyDet as the conversion electron detector to archive a better momentum resolution and higher rate capabilities. Thus, Phase-I StrECAL's secondary purpose is to serve as a prototype for the upgraded detector system.

Five straw-tracker stations (see Fig. 1) are utilized in Phase-I. Each will be made of four perpendicularly arranged planes, to precisely determine the x and y positions of passing particles. They will be deployed inside a vacuum to minimize electron multiple scattering, otherwise dominating the momentum resolution. Furthermore, the material budget for the straws themselves has to be minimized. After several R&D projects, the production of straws with only 20 µm thick aluminized mylar walls and 10 mm diameter filled with evenly mixed Argon:Ethane gas succeeded. Mass production of straw tubes for Phase-I is already finished, and even thinner straws with 12 μm walls for Phase-II are produced and under testing. A milestone achieved in 2022 is the completion of Straw Station #1. Prototype performance tests show a spatial resolution of 150 μm can be achieved [7].

The electromagnetic calorimeter is placed downstream of the five straw-tracker stations. Based on cost-performance evaluations, it was decided to use LYSO scintillator crystals, each 2 mm × 2 mm × 120 mm. Besides providing an additional hit position, it will measure the particle energy E independently from the particle momentum p obtained by the straw system. The measurement is crucial to calculate the ratio E/p used for particle identification. Results from ECAL prototype tests utilizing a vacuum chamber show an excellent energy resolution of 4 %, timing resolution of < 0.5 ns, and a 6 mm spatial resolution for 105 MeV electrons [8].

4 Conclusion

Charge lepton flavor violation is one of the most promising BSM fields currently under investigation. The combined suppression of SM and experimental background highlights the $\mu - e$ conversion search as the golden channel for discovering BSM physics in the coming decade. The COMET experiment aims to measure $\mu - e$ conversion with an unprecedented single event sensitivity of $O(10^{-15})$ and $O(10^{-17})$ in Phase-I and Phase-II, respectively. With the completion of the proton beamline leading up to the COMET experiment and successful beam tests regarding proton bunching and extinction, a first smaller test run will commence at the end of the Japanese fiscal year 2022. This so-called Phase-α will lay the foundation for the subsequent start of Phase-I. As motivated in this article, detectors for Phase-I are currently being prepared and commissioned on schedule to start the first physics data-taking in 2024.

We acknowledge support from JSPS, Japan; Belarus; NSFC, China; IHEP, China; IN2P3-CNRS, France; CC-IN2P3, France; SRNSF, Georgia; DFG, Germany; JINR; IBS, Korea; RFBR, Russia; STFC, United Kingdom; and Royal Society, United Kingdom.

References

[1] W. Altmannshofer, A.J. Buras, S. Gori, P. Paradisi, D.M. Straub, Nucl. Phys. B **830**, 17 (2010)

[2] Y. Kuno, Prog. Theor. Exp. Phys. **2013**, 022C01 (2013)

[3] W. Bertl, R. Engfer, E. Hermes, G. Kurz, T. Kozlowski, J. Kuth, G. Otter, F. Rosenbaum, N. Ryskulov, A. van der Schaaf et al., Eur. Phys. J. C **47**, 337 (2006)

[4] C.S. Wu, L. Wilets, Annu. Rev. Nucl. Part. Sci. **19**, 527 (1969)

[5] J. Heeck, R. Szafron, Y. Uesaka, Phys. Rev. D **105**, 053006 (2022)

[6] H. Nishiguchi, Y. Fujii, Y. Fukao, Y. Hashimoto, Y. Igarashi, S. Mihara, M. Moritsu, R. Muto, P. Sarin, F. Tamura et al., Extinction Measurement of J-PARC MR with 8 GeV Proton Beam for the New Muon-to-Electron Conversion Search Experiment – COMET, in Proceedings of the 10th International Particle Accelerator Conference, IPAC, 19-24 May 2019, Melbourne, Australia (2019)

[7] H. Nishiguchi, P. Evtoukhovitch, Y. Fujii, E. Hamada, S. Mihara, A. Moiseenko, K. Noguchi, K. Oishi, S. Tanaka, J. Tojo et al., Nucl. Instrum. Methods Phys. Res. Sect. A **845**, 269 (2017)

[8] K. Oishi, Development of Electromagnetic Calorimeter Using LYSO Crystals for the COMET Experiment at J-PARC, in Proceedings of the European Physical Society Conference on High Energy Physics, EPS-HEP, 5-12 July 2017, Venice, Italy (2018)

Estimated performance of the TRIUMF ultracold neutron source and electric dipole moment apparatus

Steve Sidhu[1,2,*], *Wolfgang* Schreyer[1], *Sean* Vanbergen[1,3], *Shinsuke* Kawasaki[4], *Ryohei* Matsumiya[1,5], *Takahiro* Okamura[1,3], and *Ruediger* Picker[1,2,**] for the TUCAN Collaboration

[1]TRIUMF, Vancouver, BC
[2]Simon Fraser University, Burnaby, BC
[3]University of British Columbia, Vancouver, BC
[4]KEK, Tsukuba, Ibaraki, Japan
[5]RCNP, Osaka, Ibaraki, Japan

Abstract. Searches for the permanent electric dipole moment of the neutron (nEDM) provide strong constraints on theories beyond the Standard Model of particle physics. The TUCAN collaboration is constructing a source for ultracold neutrons (UCN) and an apparatus to search for the nEDM at TRIUMF, Vancouver, Canada. In this work, we estimate that the spallation-driven UCN source based on a superfluid helium converter will provide $(1.38 \pm 0.02) \times 10^7$ polarized UCN at a density of 217 ± 3 UCN/cm^3 to a room-temperature EDM experiment per fill. With $(1.43 \pm 0.02) \times 10^6$ neutrons detected after the Ramsey cycle, the statistical sensitivity for an nEDM search per storage cycle will be $(1.94 \pm 0.06) \times 10^{-25}$ $e$cm (1σ). The goal sensitivity of 10^{-27} $e$cm (1σ) can be reached within 281 ± 16 measurement days.

1 Introduction

The TRIUMF Ultracold Advanced Neutron (TUCAN) Collaboration is building a novel ultracold neutron (UCN) source and an apparatus to search for an electric dipole moment of the neutron (nEDM). We will outline the design of source and apparatus and estimate their performance.

The nEDM is a unique observable that signifies time-reversal violation in the neutron. Assuming the combined charge, parity and time reversal (CPT) operation is symmetric, this also violates CP symmetry. So far, a non-zero nEDM has not been found. The current best upper limit is 1.8×10^{-26} $e$cm (90% CL) [1], five orders of magnitude larger than the EDM prediction by the Standard Model of particle physics [2]. Beyond-the-Standard-Model (BSM) theories tend to predict additional contributions to the nEDM. On the one hand, further improving the nEDM upper limit seriously constrains BSM theories. On the other hand, a discovery above the Standard Model prediction would be an indication of new physics or shed light on the strong CP problem. The aim of the TUCAN collaboration is to search for an nEDM at a sensitivity at the 10^{-27} e cm level (1σ).

*e-mail: ssidhu@triumf.ca
**e-mail: rpicker@triumf.ca

Similar to other experimental efforts [3–6], we will use ultracold neutrons that are characterized by their exceptionally low kinetic energies $E_{\text{kin}} < 300\,\text{neV}$. The strong interaction of neutrons with materials gives rise to Fermi's pseudopotential U_{F} and can cause reflections under all angles of incidence at a material boundary if $E_{\text{kin}} < U_{\text{F}}$. Fermi potentials can be as high as 347 neV (^{58}Ni). The interaction of the neutron's magnetic moment $\mu_{\text{n}} = -60.3$ neV/T with a strong magnetic field B gives rise to similar potentials $U_{\text{M}} = \pm\mu_{\text{n}}B$. One spin polarization is attracted by a strong magnetic field (high-field seekers, "+"), while the other is repelled (low-field seekers, "-").

The above interactions allow efficient transport of UCN through guides to an experiment, long observation times, and close to full polarization of the neutron spins, important ingredients for a highly sensitive EDM measurement. Gravity, with its potential of 102.5 neV per meter elevation also influences ultracold neutrons significantly.

2 Principles

We will now describe the working principles of the UCN source (Fig. 1) and EDM apparatus (Fig. 3) under construction.

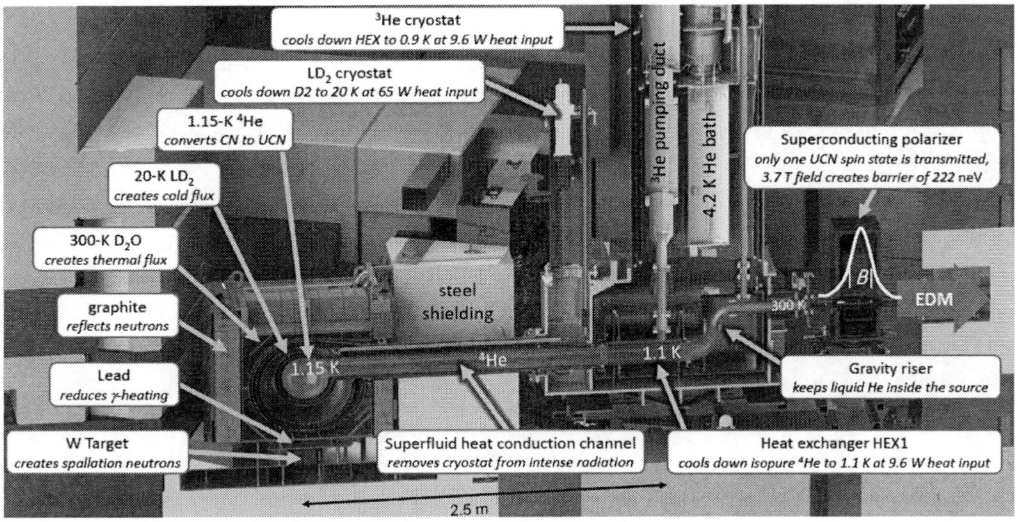

Figure 1. Cutaway view of the 3D model for the TUCAN source.

2.1 UCN production

483 MeV protons from the TRIUMF main cyclotron hit a tungsten spallation target to produce fast neutrons [7]. They are moderated in room-temperature heavy water, graphite, and liquid deuterium at 20 K, as shown in Fig. 1. These moderators surround a vessel filled with superfluid, isotopically-purified helium-4 at around 1.1 K. The geometry has been optimized to create a large flux of cold neutrons inside the helium [8]. Here, cold neutrons can excite phonons and rotons, losing almost all their energy in the process to become ultracold. We estimate a total UCN production rate of 1.4×10^7 UCN/s at a proton current of $40\,\mu$A [8]. The reverse scattering, where a UCN gains momentum by scattering on existing rotons and phonons, is suppressed by cooling the medium. Thank to this suppression and the fact that

pure helium-4 does not absorb neutrons, ultracold neutrons can be stored in the source for tens of seconds, which is similar to the time needed to fill neutrons into experiments such as the nEDM.

2.2 Cryogenics and UCN extraction

A total heat load of up to 9.6 W, 8.1 W from the spallation target and 1.5 W estimated static heat load, is deposited in the superfluid helium. This heat is transported in a helium-filled channel of 2.5 m length through steel shielding to a large helium-4 to helium-3 heat exchanger (HEX 1).

The smooth surface on the helium-4 side of HEX 1 reduces UCN losses and has an area of $0.26\,m^2$. Vertical fins increase the surface on the helium-3 side to $0.86\,m^2$. An array of pumps connected via large diameter pumping ducts lower the saturated vapor pressure above the liquid helium-3 to 6.6 mbar corresponding to a temperature of 0.89 K. The helium-3 is pre-cooled and liquefied in two liquid-helium baths and several heat exchangers [9]. The superfluid helium-4 is contained inside the source by an upwards kink in the UCN guide (gravity riser in Fig. 1).[1]

Accounting for the helium-3 boiling curve from [10, 11] and the Kapitza conductivity at the helium-4 interface we measured at KEK [12], we estimate the helium-4 temperature inside HEX1 to be 1.1 K. Liquid helium at this temperature is a two-phase fluid (He-II) with a superfluid and a normal fluid component. For heat loads as large as in our case, the heat transport is likely limited by turbulent counterflow between the two fluid components, as described by Gorter and Mellink [13]. We have calculated the resulting temperature gradient

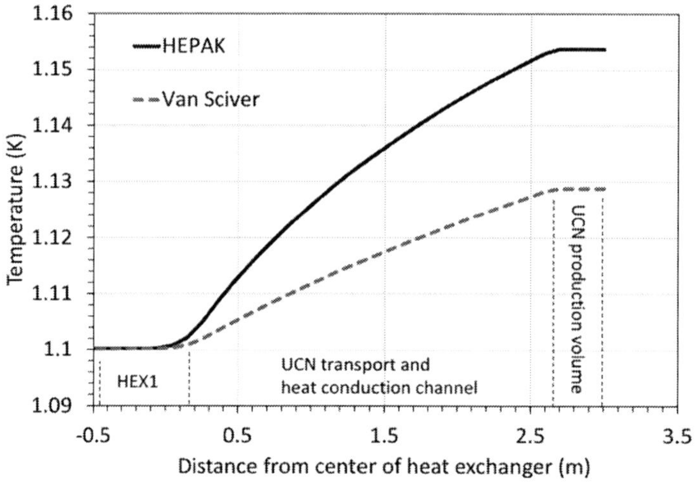

Figure 2. Temperature profile inside the He-II of the heat conduction channel in the TRIUMF UCN source according to two parametrizations of Gorter-Mellink heat transport: Van Sciver [14] and HEPAK [15]

for two parametrizations of the Gorter-Mellink heat conductivity. Results using Van Sciver's formalism [14] and data from HEPAK [15] are shown in Fig. 2. From the less favorable HEPAK data, we estimate an average upscattering lifetime for the UCN of 29 s in the source

[1] We have estimated the heat load from superfluid film flow to be less than 70 mW and therefore are not planning a knife edge or film burner in this location.

from equation $\tau_{up}^{-1}(T) = BT^7$. Our conservative estimate of $B = 0.016\,\mathrm{s^{-1}K^{-7}}$ stems from prior measurements [16–18] and our own extraction based on experiments with a prototype UCN source using superfluid helium [19]. We only account for two-phonon scattering and assume the single-phonon and roton-phonon scattering contributions are negligible. Upscattering in the helium vapour above the liquid has a minor impact, reducing the lifetime of the neutrons by around 0.1 s.

UCN are extracted through an aluminum foil ($U_{Al} = 54\,\mathrm{neV}$) at room temperature separating the helium-vapour-filled region above the gravity riser from the vacuum guides to the experiment. It is mounted inside the warm bore of a superconducting magnet that provides a 3.7 T field ($U_M = 222\,\mathrm{neV}$). This accelerates high-field-seeking UCN through the foil ($U_{Al} - U_M = -168\,\mathrm{neV}$) and reflects low-field-seeking UCN below $U_{Al} + U_M = 276\,\mathrm{neV}$, creating a fully polarized UCN flow to the EDM experiment.

Cooling of the liquid-deuterium moderator to around 20 K at a heat load of 65 W from the spallation target is provided by a Gifford-McMahon cryo-cooler inside the LD_2 cryostat. It is connected to the deuterium moderator volume via a thermosyphon loop.

2.3 EDM experiment

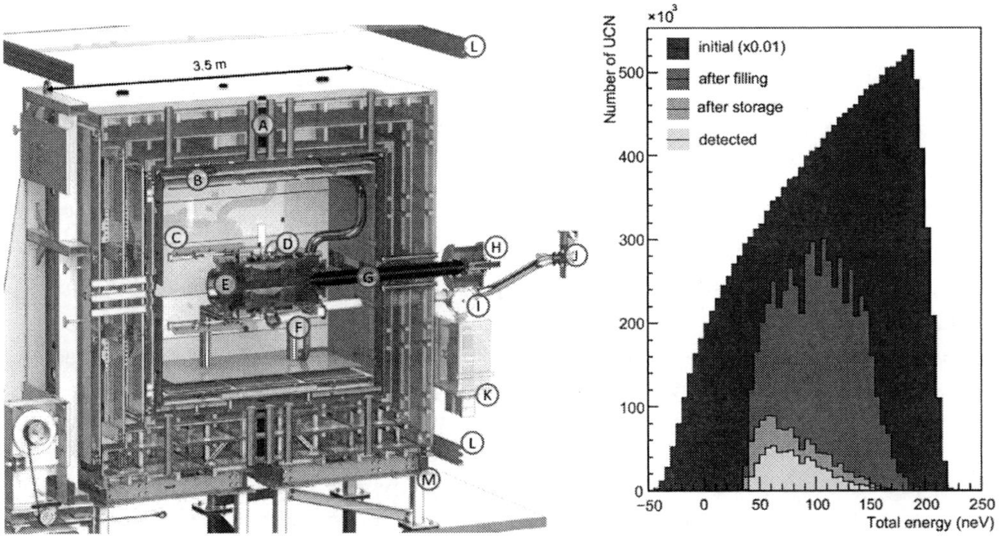

Figure 3. *Left*: Cutaway view of the 3D model for the TUCAN EDM Apparatus. (A) 4-layer magnetically shielded room, (B) Self-shielded B_0 coil, (C) EDM cell valve actuation, (D) EDM cell shutter, (E) Central high-voltage electrode, (F) UCN guide, (G) High-voltage feeder, (H) High-voltage vacuum feed through, (I) EDM detector switch, (J) UCN and vacuum valve, (K) Spin analyzers and detectors, (L) Coil compensating the ambient magnetic field, (M) Support structure. *Right*: Spectra of kinetic energy plus gravitational potential energy of UCNs determined by Monte Carlo simulations using PENTrack [20]. The initial spectrum produced in the source is scaled by a factor of 0.01 in the plot.

The TUCAN EDM experiment will use Ramsey's technique of separated oscillating fields [21] in a double-chamber EDM spectrometer at room temperature as shown in Fig. 3. This is comparable to other current EDM experimental efforts [3–5]. Polarized ultracold neutrons will be filled into two vertically stacked EDM cells, each with a height of 0.16 m and a diameter of 0.5 m. A central high-voltage electrode at 200 kV in between the two cells and two ground electrodes create vertical electric fields with average magnitudes of $E = \pm 12.5\,\text{kV/cm}$ in top and bottom cell, respectively.[2] Insulating rings coated with deuterated polystyrene (dPS) form the outer wall between the electrodes. Non-magnetic, linearly actuated shutters complete the EDM cells.

The required homogeneous and stable magnetic field environment is created by a magnetically shielded room with 4 layers of mu-metal for DC and one copper layer for AC shielding. The estimated DC shielding factor is $\approx 10^5$. A self-shielded coil and several shim coils create a uniform, vertical magnetic field on the order of $B_0 = 1\,\mu\text{T}$. We will introduce polarized mercury vapour into the EDM cells to probe the magnetic field in the same volume that the UCN occupy. The $\pi/2$ spin flip required for both UCN and Hg will be initiated by a magnetic field of roughly 10 nT that oscillates close to the Larmor frequencies of either species.

After the Ramsey cycle, four Li-glass scintillation detectors [22], two connected to each EDM cell, count the neutrons. A spin flipper and a spin analyzer using a magnetized iron film coated onto an aluminum foil are placed on top of each detector. Low field seekers are not able to penetrate the iron film ($U_{\text{Fe}} = 209\,\text{neV}$) since the potential barrier is increased to $U_{\text{Fe}} - \mu_n B_{\text{Fe}} = 329.6\,\text{neV}$ by the magnetic field $B_{\text{Fe}} = 2\,\text{T}$ inside the saturated iron film. High-field seekers only see a barrier of $U_{\text{Fe}} + \mu_n B_{\text{Fe}} = 88.4\,\text{neV}$ that is overcome by placing the analyzers low enough so that the UCN gain more energy in the gravitational fields. The spin-flipping AC magnetic fields are only activated on one arm per EDM cell, therefore allowing simultaneous counting of both polarization states of the UCN for both cells.

3 Performance estimation

To estimate and improve the UCN source and EDM experiment performance we studied its UCN particle and spin transport characteristics in great detail. Many factors have an influence.

Absorption and upscattering on and in materials can be expressed via an imaginary Fermi potential $U_F = V_F - iW_F$. The larger the real part V_F of wall materials, the larger the energy spread of UCN may be; the larger the imaginary part W_F the higher the loss probability per wall bounce, see [20].

The **geometry** of UCN storage vessels and guides influences the mean free path between wall hits and therefore the losses of UCN due to W_F. Any elevation change of UCN transport components leads to a kinetic energy change; some neutrons might not reach higher elevations.

The majority of UCN wall bounces is specular, but roughness of materials causes **non-specular reflection**. The simplest model called Lambert reflection assumes a cosine distribution of the outgoing neutron directions around the surface normal for any angle of incidence [23]. In this model, the Lambert reflection probability P_L is the sole input parameter.

Wall interactions of UCN can also cause **spin flips**, important for any experiment requiring polarized neutrons. This can be expressed by a depolarization probability per wall bounce β.

Magnetic fields influence UCN trajectories via forces in magnetic gradient fields. The precession of neutron spins in magnetic fields is a very important aspect for the neutron EDM experiment.

All of these interactions of ultracold neutrons are included in the Monte Carlo simulation package PENTrack [20]. PENTrack performs spatial tracking of UCN in gravitational and magnetic fields taking into account surface interactions described above as well as spin tracking in magnetic and electric fields. We used this package for extensive studies of many aspects of the TUCAN components and their influence on the yield of the UCN source and the sensitivity of the EDM experiment [24].

[2]Our simulations conservatively assume NiP coated electrodes, but we intend to use diamond-like carbon (DLC) coating due to its better performance.

Table 1. Selected input parameters for UCN Monte Carlo simulations and the sensitivity calculation.

Parameter	Value	Source
Kinetic energy spectrum produced in source	$\propto \sqrt{E}$, Fig. 3	Low-energy tail of Maxwell spectrum
Base temperature of superfluid helium at HEX1	1.1 K	Cryogenic calculations of cryostat [9], 3He boiling curve [10, 11], Kapitza resistance [12]
Temperature gradient in superfluid helium	Fig. 2	Gorter-Mellink formalism [14, 15]
Upscattering lifetime in liquid helium $\tau_{\text{up}}^{-1}(T) = BT^7$	$B = 0.016\,1/\text{s/K}^7$	[16–19]
Fermi potential of		
- superfluid helium	$19\,\text{neV} - i\frac{\hbar}{2}BT^7$	[16–19]
- NiP coated components	$213 - i0.07\,\text{neV}$	UCN experiments at TRIUMF,
- dPS insulator ring	$171 - i0.047\,\text{neV}$	J-Parc, and LANL [25, 26]
Lambert diffuse reflection probability in		
- cryogenic region	$P_L = 0.15$	Roughness measurements
- room temperature guides	$P_L = 0.03$	UCN expt's at TRIUMF [25]
Spin-flip prob. per wall bounce	3×10^{-5}	10× higher than [27–29]
Avg. electric field in EDM cells	12.5 kV/cm	Finite element simulations
Total spin coherence lifetime T_2	$> 800\ s$	Estimate based on [1, 30]

3.1 Input parameters and simulation procedure

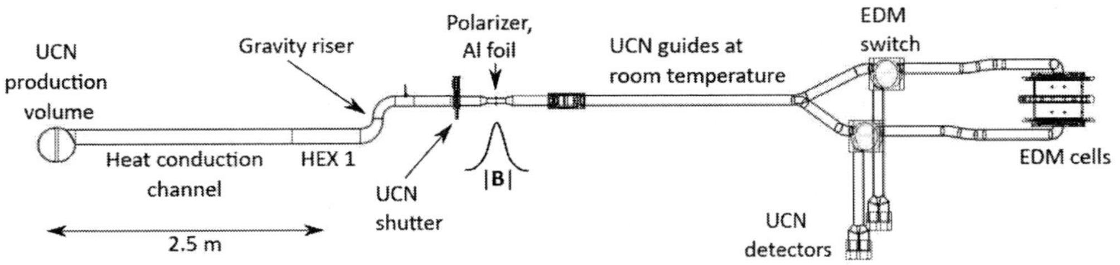

Figure 4. Side view of one exemplary geometry used to simulate the TUCAN source and EDM experiment.

Tab. 1 provides a summary of input parameters used in our simulations and calculations. We have run sensitivity studies on most parameters and on a lot of geometrical dimensions of the UCN source and EDM experiment. These studies can be found in [24].

In this work, due to space constraints, we concentrate on the UCN source as it is being built and the EDM experiment as currently planned, see Figs. 1 and 3. The goal is to estimate the required measurement time to achieve a statistical sensitivity of $10^{-27}\ e$cm (1σ).

In the simulation, we separate the EDM experimental cycle into four major stages. (1) UCN production in the source: For a duration T_P UCN are produced with a homogeneous spatial distribution and uniform starting angle distribution inside the UCN production volume, see Fig. 4. The UCN shutter is closed. (2) UCN production and filling of the EDM cells: the UCN shutter and EDM cell shutters are opened, UCN production continues and the neutrons are allowed to diffuse into the EDM cells for a duration T_F. The spins are filtered

by the magnetic field of the polarizer. (3) UCN storage and Ramsey period: UCN production stops, the EDM cell shutter is closed, UCN are stored inside the cells. A first spin flip starts free precession of the UCN spins. After the free precession time T_R, a second spin flip turns any change of the Larmor precession frequency into a polarization change (Ramsey method). (4) UCN detection and spin analysis: The EDM cell shutter is opened, UCN can diffuse to spin-sensitive detectors for a duration T_{det}, allowing for a polarization measurement.
This separation allows us to find the durations for all stages that minimize the measurement time required to reach the desired sensitivity, see [24].

The magnetic field of the superconducting polarizer is included in the simulation as a 3D field map, created by an Opera finite-element simulation [31] and cross checked by magnetic field measurements of the actual magnet. The spin analyzers and detectors are realized slightly differently in our simulation model. Spin flippers are represented by volumes inside the UCN guides with a spin flip probability of 99%, our anticipated spin-flipper efficiency. The highly localized magnetic field inside the very thin, iron-coated aluminum foil of a spin analyzer cannot be represented via a field map. The magnetization is instead represented by a spin-dependent pseudo potential inside the foil as described at the end of Sec. 2.3.

To determine the sensitivity reach of the EDM experiment we use the well-known sensitivity formula for Ramsey-type EDM experiments $\sigma(d_n) \geq \frac{\hbar}{2E\alpha_{det}T_R\sqrt{N_{det}}}$. E denotes the electric field, α_{det} the visibility of the Ramsey pattern, mainly driven by polarization, depolarization and spin coherence. T_R is the free precision time and N_{det} the number of UCN detected per Ramsey cycle.

We expect to have a magnetically quiet-enough environment at TRIUMF to conduct the EDM experiment outside business hours (18:00 to 7:00 on weekdays and the whole weekend). Our figure of merit is the number of days required to reach a statistical sensitivity of 10^{-27} $e$cm (1σ) when taking data during these hours.

3.2 Results

Extensive simulation campaigns have been performed to understand and improve the performance of the TUCAN source and EDM experiment. The dependence of the results on the assumptions made, especially material properties, were studied. We will mainly focus on dependence on three parameters very relevant to UCN storage and transport that are notoriously hard to measure: the imaginary Fermi potential, the diffuse reflection probability in the Lambert model and the spin flip probability on wall interaction.

In the simulation we assume all UCN exposed surfaces to be coated with electro-less nickel (NiP) with a high phosphorous content above 10%, except the Al foil inside the polarizer (54.1 – i0.00281 neV), the Al analyzer foil coated with iron (209.1 – i0.000872 neV) and the EDM cell insulators coated with dPS (171 – i0.047 neV). Our measurements indicate a range for the NiP imaginary Fermi potential of roughly [0.03,0.09] neV [26]. Fig. 5 (top left) indicates that the required measurement time to reach 10^{-27} $e$cm varies by a factor of three mainly driven by the UCN storage lifetime in the EDM cells.

The diffuse reflection probability mainly impedes UCN transport through guides. Varying it from 1 % to 16 % changes the number of neutrons loaded into the EDM cells and the measurement time dramatically, see Fig. 5 (right top).

For spin-sensitive experiments like the EDM, spin transport properties are very important. We set the baseline probability for spin flip on wall bounce on the NiP surfaces to a very conservative 3×10^{-5}, 10 times higher than literature values [27–29]. A parameter scan in our simulation shows lower depolarization at the walls would reduce the measurement time and improve the visibility α_{det} of the Ramsey fringes significantly.

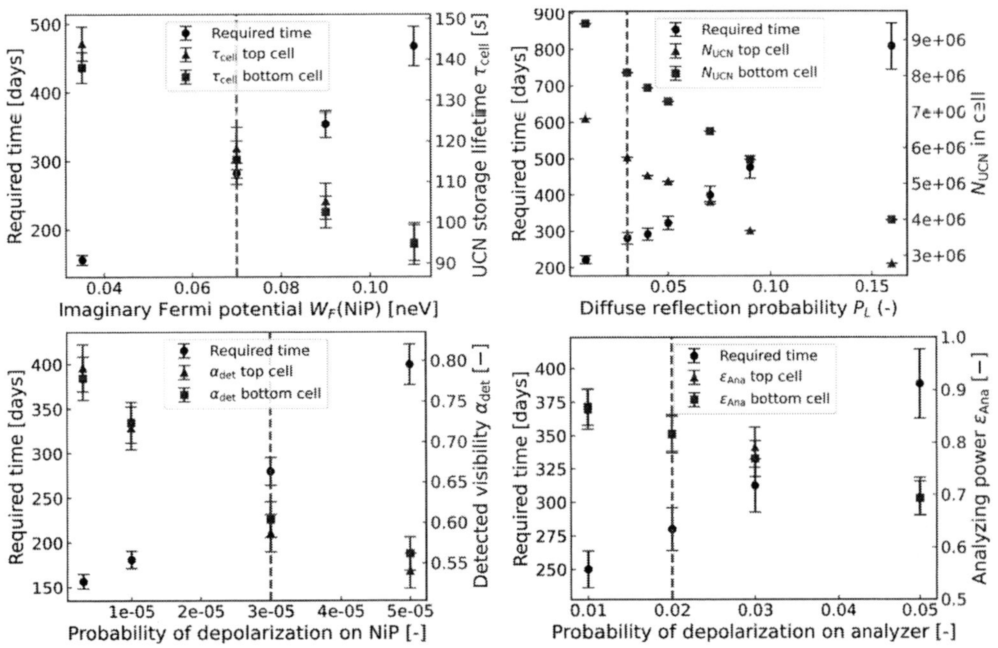

Figure 5. Four PENTrack sensitivity studies showing the required measurement time to reach a statistical sensitivity of 10^{-27} $e$cm (1σ) and a selected other parameter for each plot. Our parameter assumptions are indicated by the vertical green lines. For more information, please refer to the text.

Additional depolarization is expected at the iron coated analyzer foil. Varying it from 1% to 5% has a rather mild effect on the required measurement time and changes the analyzing power of the spin sensitive detection by around 25%.

We performed numerous simulation studies varying geometrical parameters of source and experiment, leading to the design as shown in Fig. 4 [24]. The optimum experiment timings are: $T_P = 20$ s, $T_F = 105$ s, $T_R = 188$ s, and $T_{det} = 49$ s. The UCN storage lifetime inside the source, an important benchmark parameter, is estimated to be 19.0 ± 0.2 s, mainly dominated by the upscattering lifetime inside the liquid helium. The expected storage lifetimes inside the top and bottom EDM cells are 119 ± 4 s and 116 ± 4 s, respectively. Fig. 3 shows the initial spectrum of kinetic plus gravitational potential energy as well as spectra right after stage (2) filling, (3) storage and for the detected UCN. Higher-energy UCN are less likely to survive and to be detected, favoring an optimization towards lower energy UCN.

For our conservative baseline assumptions, we expect to fill $(1.38 \pm 0.02) \times 10^7$ neutrons into the EDM experiment, which corresponds to a density of $217 \pm 3\,\mathrm{UCN/cm^3}$, and detect $N_D = (1.43 \pm 0.02) \times 10^6$ after the Ramsey cycle assuming a detector efficiency of 90%. The initial UCN polarization in the EDM cell is projected to be (92 ± 2)% taking into account the depolarization per bounce from Tab. 1. This reduces to (73 ± 2)% after the Ramsey cycle, corresponding to a spin coherence time $T_2 \approx 830$ s. The detected visibility is estimated to be $\alpha_{det} = (60 \pm 2)$%. The sensitivity per cycle then becomes $(1.94 \pm 0.06) \times 10^{-25}$ $e$cm.

This should allow us to achieve a statistical sensitivity of 10^{-27} $e$cm (1σ) in 281 ± 16 calendar days of data taking. Quoted uncertainties are calculated from the simulation statistics and fitting errors. We will perform commissioning experiments to validate the input parameters for our calculations as we build the source and experiment components.

4 Conclusion

The TUCAN collaboration is building an ultracold neutron source based on a spallation target and a superfluid helium converter as well as a room-temperature apparatus to search for the electric dipole moment of the neutron at TRIUMF. Extensive simulation campaigns have been performed to understand and improve source and experiment performance. Their results have driven the designs of UCN source and EDM apparatus as shown above.

We estimate to reach a statistical sensitivity of 10^{-27} $e$cm (1σ) in 281 ± 16 calendar days or less than three years of data taking.

References

[1] C. Abel, S. Afach, N.J. Ayres et al., Phys. Rev. Lett. **124**, 081803 (2020)
[2] B.H. McKellar, S. Choudhury, X.G. He et al., Phys. Lett. B **197**, 556 (1987)
[3] D. Wurm, Ph.D. thesis, Universität München (2021)
[4] N.J. Ayres, G. Ban, L. Bienstman et al., Eur. Phys. J. C **81**, 1 (2021)
[5] T.M. Ito, E.R. Adamek, N.B. Callahan et al., Phys. Rev. C **97**, 012501 (2018)
[6] E.P. Tsentalovich (nEDM), Phys. Part. Nucl. **45**, 249 (2014)
[7] S. Ahmed, T. Andalib, M. Barnes et al., Nucl. Instr. Meth. A **927**, 101 (2019)
[8] W. Schreyer, C. Davis, S. Kawasaki et al., Nucl. Instr. Meth. A **959**, 163525 (2020)
[9] T. Okamura, S. Kawasaki, IOP Conf. Ser.: Mater. Sci. Eng. **755**, 012141 (2020)
[10] M. Maeda, A. Beppu, Y. Fujii et al., Cryogenics **40**, 713 (2000)
[11] M. Tanaka, T. Kodama, Cryogenics **29**, 203 (1989)
[12] Kawasaki, Shinsuke, Okamura, Takahiro, EPJ Web Conf. **219**, 10001 (2019)
[13] C. Gorter, J. Mellink, Physica **15**, 285 (1949)
[14] S.W. Van Sciver, *Helium cryogenics* (Springer Science & Business Media, 2012)
[15] Cryodata.Inc., *HEPAK User's Guide* (Horizon Technologies, 2005), `https://htess.com/hepak/`
[16] K. Leung, S. Ivanov, F.M. Piegsa et al., Phys. Rev. C **93**, 025501 (2016)
[17] R. Golub, C. Jewell, P. Ageron et al., Z. Phys. B **51**, 187 (1983)
[18] H. Yoshiki, K. Sakai, M. Ogura et al., Phys. Rev. Lett. **68**, 1323 (1992)
[19] S. Vanbergen et al. (TUCAN Collaboration), Phys. Rev. C in preparation (2023)
[20] W. Schreyer, T. Kikawa, M.J. Losekamm et al., Nucl. Instr. Meth. A **858**, 123 (2017)
[21] N.F. Ramsey, Phys. Rev. **78**, 695 (1950)
[22] Jamieson, B., Rebenitsch, L. A., Hansen-Romu, S. et al., Eur. Phys. J. A **53**, 3 (2017)
[23] J.H. Lambert, *Photometria sive de mensura et gradibus luminis, colorum et umbrae* (Klett, 1760)
[24] S. Sidhu, Ph.D. thesis, to be submitted to Simon Fraser University (2023)
[25] S. Ahmed, E. Altiere, T. Andalib et al., Phys. Rev. C **99**, 025503 (2019)
[26] H. Akatsuka, T. Andalib, B. Bell et al., submitted to Nucl. Instr. Meth. A (2022)
[27] Z. Tang et al., Nucl. Instrum. Meth. A **827**, 32 (2016), `1510.06490`
[28] T. Brys, M. Daum, P. Fierlinger et al., J. Res. Natl. Inst **110**, 279 (2005)
[29] V. Bondar, S. Chesnevskaya, M. Daum et al., Phys. Rev. C **96**, 035205 (2017)
[30] C.A. Baker, D.D. Doyle, P. Geltenbort et al., Phys. Rev. Lett. **97**, 131801 (2006)
[31] Dassault Systems, *Opera*, Nov 14, 2022, `www.3ds.com/products-services/simulia/products/opera/`

New beamlines and future prospects of the J-PARC muon facility

Takayuki Yamazaki[1*], *Naritoshi* Kawamura[1], *Koichiro* Shimomura[1], *Yasuhiro* Miyake[1], *Yu* Oishi[1], *Taihei* Adachi[1], *Patrick* Strasser[1], *Akihiro* Koda[1], *Hiroshi* Fujimori[1], *Takahiro* Yuasa[1], *Yutaka* Ikedo[1], *Yasuo* Kobayashi[1], *Ken-ichi* Sasaki[1], *Tsutomu* Mibe[1], *Takahiro* Ushizawa[2], *Yuta* Higashino[3], *Daiki* Nagao[3], *Masaharu* Aoki[3], *Kazuya* Hase[4], *Satoshi* Kaneko[4], *Ryota* Tagawa[4], *Taichi* Uematsu[4], *Yoshihiro* Seiya[4], *Satoshi* Uetake[5], *Takahiro* Masuda[5], *Hiroki* Tada[6], *So* Sugiyama[6], and *Kazuhito* Suzuki[6]

[1]High Energy Accelerator Research Organization (KEK), Ibaraki 305-0801, Japan
[2]The Graduate University for Advanced Studies (SOKENDAI), Ibaraki 305-0801, Japan
[3]Osaka University, Osaka 560-0043, Japan
[4]Osaka Metropolitan University, Osaka 558-8585, Japan
[5]Okayama University, Okayama 700-8530, Japan
[6]Nagoya University, Nagoya 464-8601, Japan

Abstract. At the J-PARC muon facility (MUSE), new beamlines started operation recently. H-line is a high-intensity pulsed muon beamline for fundamental physics experiments. The first beam of the H-line was delivered to its first branch (H1 area) in January 2022, where a precise measurement of the muonium hyperfine structure and a search for μ-e conversion will be conducted. Further extension of the second branch of the H-line for a muon g-2/EDM experiment and a transmission muon microscope project is also ongoing. In addition, the second branch of the surface muon beamline (S2 area of the S-line) was opened for a muonium 1S-2S spectroscopy experiment in FY2021. In this paper, the recent upgrade and present status of the J-PARC muon facility and its prospects are presented.

1 Introduction

The J-PARC muon facility (MUSE) is a pulsed muon beam facility in the Material and Life Science Facility (MLF) of the J-PARC. The 3 GeV proton beam from the RCS hits a rotating graphite target and produces charged pions via proton-nucleon reactions. Muons produced by the decay of charged pions are captured and delivered to experimental areas using electromagnets. There are four muon beamlines and seven experimental areas at MUSE. Recently, one new beamline (H-line) and two new experimental areas (H1 and S2 area) for fundamental physics programs started operation.

2 New beamlines for fundamental physics

* Corresponding author: takayuki@post.kek.jp

2.1 H-line

H-line is a muon beamline in the east #1 experimental hall of MLF. It is a general-purpose beamline [1] that can generate both positive and negative muons. The capture electromagnet of the H-line has a large acceptance of 108 mSr, which enables us to deliver high-intense muon beams. The design intensity of surface muons at the H-line is 10^8 muons/s with a proton beam power of 1 MW. An experimental area at the first branch of the H-line named H1 area started operation in January 2022, where the hyperfine splitting measurement of muonium (MuSEUM experiment [2]) and the search for a cLFV (charged Lepton Flavour Violation) process (DeeMe experiment [3]) are planned. Further extension of the second branch of the H-line is ongoing, at which a low emittance muon beam will be produced by re-accelerating ultra-slow muons to be used for the muon g-2/EDM experiment at J-PARC [4] and a transmission muon microscope (TμM) project.

2.1.1 Beam commissioning at H1 area

Beam commissioning of the H-line was conducted at the H1 area. Because a DC separator (Wien filter) as a velocity selector is not installed in the H-line yet, a large positron/electron background existed during the beam commissioning. It made the beam commissioning difficult, but we managed to measure the beam intensity, profile, and momentum of surface muons and the beam intensities of negative muons at several momenta. Installation of a DC separator is scheduled within JFY2022.

Figure 1 shows the setup to measure the intensity of surface muons. Decay positrons from the muon-stopping target are measured using a pair of plastic scintillators. The number of detected positrons was obtained by fitting the time spectrum of decay positrons shown in Fig. 2. To estimate the muon beam intensity in the H-line, we estimated the detector acceptance and the transmission efficiency of the Kapton window using a Geant4-based simulation [5] and measured the detection efficiency of the plastic scintillator pair. As a preliminary result, the surface muon beam intensity was estimated to be $(9.9\pm1.0)\times10^7$ muon/s for 1 MW proton beam.

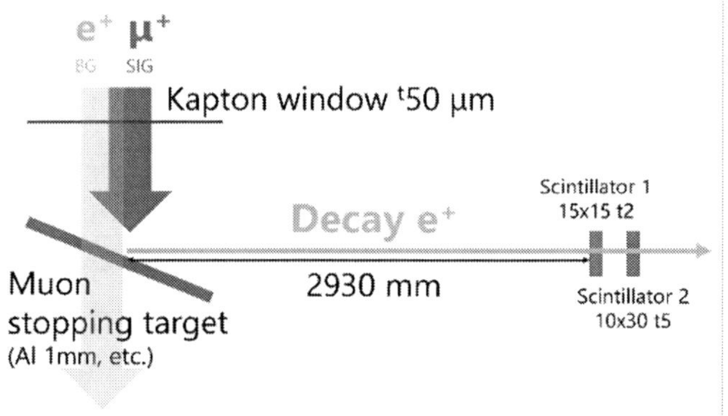

Fig. 1. Experimental setup to measure the intensity of surface muons

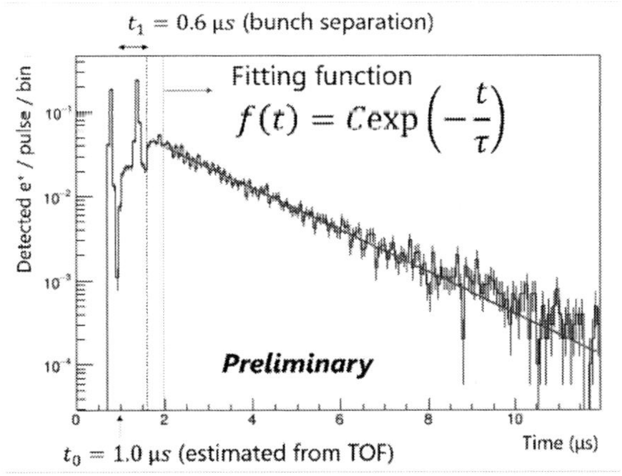

Fig. 2. Time spectrum of decay positrons in the intensity measurement of surface muons

The momentum of surface muons was also measured with almost the same setup as the intensity measurement, but a magnetic field of 23.85 G was applied to an Aluminium target of various thicknesses. Due to the muon spin rotation in a magnetic field, the time spectrum of decay positrons showed a wiggled structure. In Fig. 3, the beat amplitudes are plotted as a function of the thickness of Al targets. The range of surface muons in Al was obtained by fitting this plot with an error function, and then we estimated the beam momentum to be 28.0 MeV/c (RMS 1.2 MeV/c) as expected.

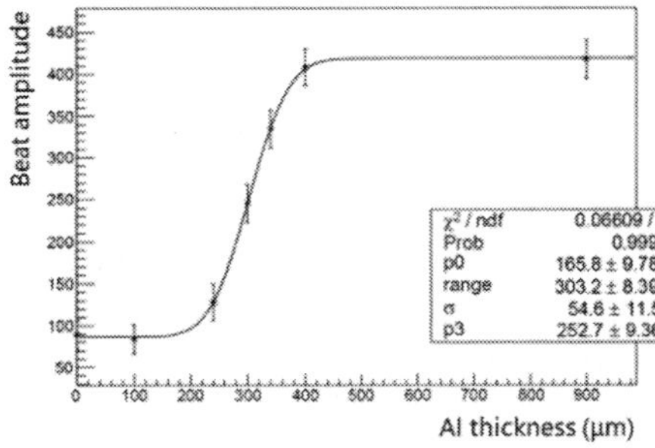

Fig. 3. Range measurement of surface muons to estimate their momenta

The beam profile of surface muons at the H1 area was measured using a beam profile monitor which is developed by T. U. Ito, *et al.* [6]. A typical beam profile is shown in Fig. 4, whose horizontal and vertical sizes are $\sigma_x = 44$ mm and $\sigma_y = 24$ mm, respectively.

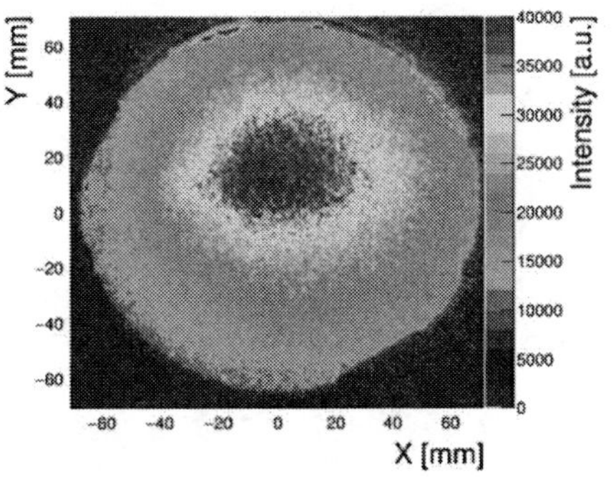

Fig. 4. Typical beam profile of surface muons at the H1 area

As the commissioning of negative muons, beam intensities with several momentum setups were measured. The measured intensities have good agreement with a Geant4-based simulation as shown in Fig. 5. Due to the failure of the power supplies of the muon/pion capture solenoid of the H-line, a plateau is observed above 50 MeV/c. The power supplies are going to be fixed within JFY2022, and then we will be able to deliver more negative muons with higher momentum at the H1 area.

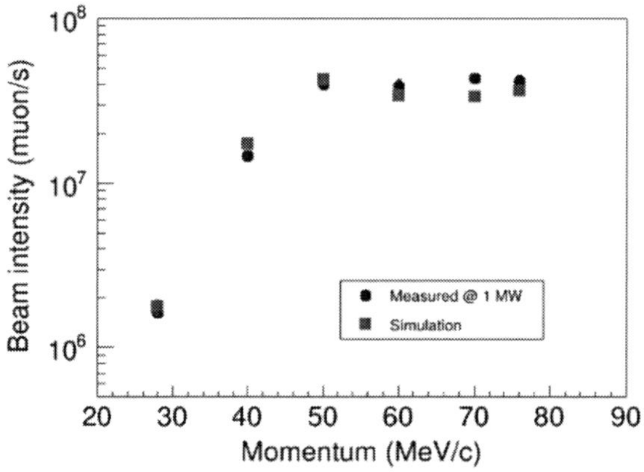

Fig. 5. Intensity of negative muons as a function of momentum of muons

2.1.2 Extension of the H-line

The extension of the second branch of the H-line is ongoing. The second experimental area named H2 area is under construction inside the MLF, where ultra-slow muons are produced and re-accelerated up to 4.5 MeV as a first stage to produce a low emittance muon beam. Radiation shields of the H2 area were already fabricated and installed, and the first beam at the H2 area is scheduled within JFY2022.

Further extension is planned in a new building named H-line experimental building as shown in Fig.6. A low-emittance muon beam accelerated up to 212 MeV will be utilized to precisely measure the muon *g-2*/EDM. In the H-line experimental building, a transmission muon microscope (TμM) is also planned to be installed, which enables non-destructive observation of thick samples. Preparatory work before starting construction of the new building, such as geotechnical investigation, a survey of the construction site, the relocation of existing cables, and so on, is ongoing.

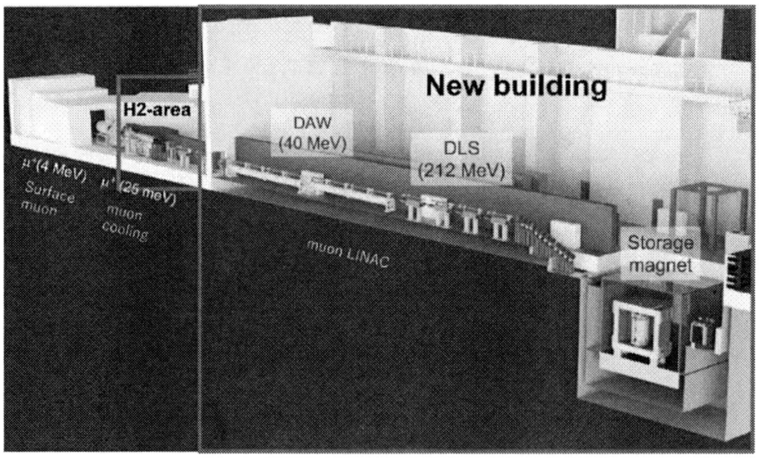

Fig. 6. Extension plan of the H-line.

2.2 S2 area of S-line

The S-line is a surface muon beamline that has two branches named S1 and S2 area. By using a kicker device installed in the S-line, the S1 and S2 areas can be operated simultaneously. The S2 area started its operation in 2021, where the spectroscopy of the 1S-2S interval of muonium is underway to determine the muonium mass precisely.

3 Second target station

A new target station (second target station, "TS2") to produce muons and neutrons is under consideration as one of the future programs of J-PARC MLF. The proposed construction site is on the west side of the MLF. At TS2, a common target is utilized to produce both muons and neutrons. Compared to the current muon production target in the MLF, which is in tandem with the neutron target, ten times more muons are produced in the target of TS2. In addition, the extraction efficiency of muons will be about 5 times better by placing a large-bore capture solenoid in the vicinity of the target. Therefore, more than a 50-fold increase in muon beam intensity is expected.

4 Summary

A new beamline (H-line) and two new experimental areas (H1 and S2 area) began operation recently. H-line is a high-intense muon beamline and surface muons of about 10^8 muons/s are obtained at the H1 area. The momentum and profile of surface muons were also measured and are consistent with our expectations. Negative muon yields at high momentum are not as high as expected because the pion/muon capture solenoid of the H-line cannot be fully excited due to the failure of its power supplies. They are scheduled to be repaired within JFY2022.

Further extension of the H-line is underway to produce a novel low-emittance muon beam by re-accelerating ultra-slow muons. H2 area in the MLF is being constructed with the aim of starting operation within JFY2022. Preparatory work to build a new building for the muon *g-2*/EDM experiment and the transmission muon microscope is underway.

The S2 area in the second branch of the S-line started operation in 2021 and a precise measurement of the mass of muonium is ongoing at the S2 area.

The conceptual design of the second target station (TS2) is in progress. Its greatest advantage is the increase in the muon beam intensity. In TS2, neutrons and muons are produced in the same target and a muon/pion capture solenoid will be placed in the vicinity of the target. These characteristics enable a more than 50-fold increase in muon yields.

This work is supported by JSPS KAKENHI Grant Number JP19H05606.

References

1. N. Kawamura, et al., Prog. Theor. Exp. Phys. **2018**, 113G01 (2018)
2. K. Shimomura, AIP conf. proc. **1382**, 245 (2011)
3. Y. Nakatsugawa, on behalf of the DeeMe Collaboration, Nucl. Part. Phys. Proc. **273-275**, 1692-1698 (2016)
4. M. Abe, et al., Prog. Theor. Exp. Phys. **2019**, 053C02 (2019)
5. S. Agostinelli, et al., Nucl. Instrum. Methods Phys. Res. A **506**, 250-303 (2003)
6. T.U. Ito, et al., Nucl. Instrum. Methods Phys. Res. A **754**, 1-9 (2014)

Towards an electrostatic storage ring for fundamental physics measurements

Chiara Brandenstein[1,*], *Stefan* Stelzl[1,2,**], *Erwin* Gutsmiedl[1], *Wolfgang* Schott[1], *Andreas* Weiler[1], and *Peter* Fierlinger[1]

[1]Technical University of Munich, Physics Department
[2]École Polytechnique Fédérale de Lausanne, Theoretical Particle Physics Laboratory

Abstract. We describe a new table-top electrostatic storage ring concept for 30 keV polarized ions with fixed spin orientation. The device will ultimately be capable of measuring magnetic fields with a resolution of 10^{-20} T with sub-mHz bandwidth. With the possibility to store different kinds of ions or ionic molecules and access to prepare and probe states of the systems using lasers and SQUIDs, it can be used to search for electric dipole moments (EDMs) of electrons and nucleons, as well as axion-like particle dark matter and dark photon dark matter. Its sensitivity potential stems from several hours of storage time, comparably long spin coherence times, and the possibility to trap up to 10^9 particles in bunches with possibly different state preparations for differential measurements. As a dark matter experiment, it is most sensitive in the mass range of 10^{-10} to 10^{-19} eV, where it can potentially probe couplings orders of magnitude below current and proposed laboratory experiments.

1 Introduction

The axion, a light pseudoscalar particle, has gained increasing attention due to a growing experimental program and new theoretical ideas. The axion could constitute a significant fraction of dark matter. A spin-dependent coupling to ordinary matter then leads to the novel signature of an effective oscillating magnetic field that can be detected using nuclear magnetic resonance techniques, see e.g., [1–4]. Related proposals that can test the axion couplings to matter can be found in [5–25], some of which [5–9] in the context of storage rings. Furthermore, we have yet to find hints for additional sources of CP violation in EDM measurements [26]. Here, storage ring precision experiments gain increasing attention as they provide alternative access via extreme measurement precision.
In this paper, we describe a versatile, table-top-scale electrostatic storage ring with 5.7 m circumference for non-relativistic, polarized ions, built to search for electric dipole moments of fundamental systems (EDMs) and to probe magnetic field-like effects resulting from ultra-light dark matter such as axion-like particles (ALPs) and dark photons. Furthermore, a combination of the device with a large volume nuclear spin polarized sample is proposed to increase the sensitivity for the ALP dark matter and dark photon dark matter measurement.

*e-mail: chiara.brandenstein@tum.de
**e-mail: stefan.stelzl@epfl.ch

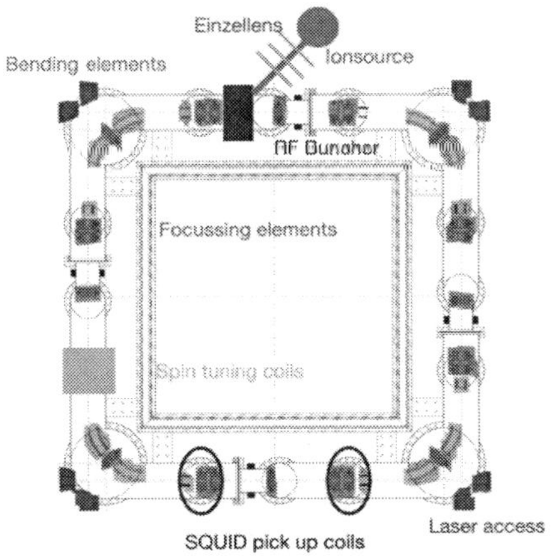

Figure 1. Sketch of the storage ring with bending elements, focussing elements, SQUIDs, spin tuning coils and RF buncher.

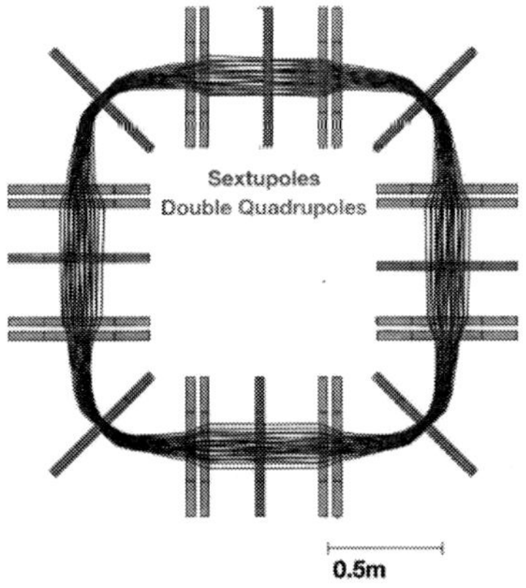

Figure 2. Beam stability simulation with sextupoles and double quadrupoles.

2 The storage ring

The experimental setup comprises an all-electrostatic storage ring for ions and ionic molecules. The demonstrator setup can store up to 10^9 charged particles in 200 bunches, confined and tuned by electrostatic potentials on periodic trajectories. However, the amount of particles is limited due to space charge effects [27], where we assume that the storage of 10^7ions/(mm · mrad) is possible [28]. The whole ring is kept at ultra-high vacuum conditions inside a stainless steel housing to enable extremely long trapping times of up to 5 h for the cryogenic setup. Long trapping times for 10^7 ions were already demonstrated at the CSR in Heidelberg [29], while long spin coherence times have been demonstrated at the COSY ring [30]. Access for spectroscopy lasers allows to manipulate and monitor the polarization of the particles during storage, enabling the use of the device as a magnetometer. Through electrostatic optics and radio frequency elements, the particles are bunched and focused such that polarization can be read out either optically or with SQUID magnetometers. We plan on doing differential measurements between flipped reference bunches, to significantly reduce the systematics.

3 Experimental Setup

(i) *The vacuum system and ion optics* for the initial implementation will be a demonstrator setup, operated at room temperature, while the final stage will be at cryogenic conditions, similar to the CSR in Heidelberg [29]. The chamber design was adapted from the CSR and comprises deflectors to bend in two steps (6 ° and 39 °), as well as sextupoles and double quadrupoles for focussing and beam stability reasons as shown in Fig.1 and 2. The sextupoles were added to compensate for chromaticity. For the demonstrator setup, we first envisage trapping 30 keV barium ions at 36 kHz. The beam stability for these ions was checked with Gicosy [31], see Fig. 2.

(ii) *The ion source* is a laser ablation source required to produce 10^6 ions in 100 ns packets, which are then electrostatically accelerated and injected into the ring. With the laser ablation source, small spot sizes of about 100 μm or smaller are possible [32]. The spot size is monitored with a multi-channel plate (see Fig. 3 for a sketch of the ion source). To demonstrate the

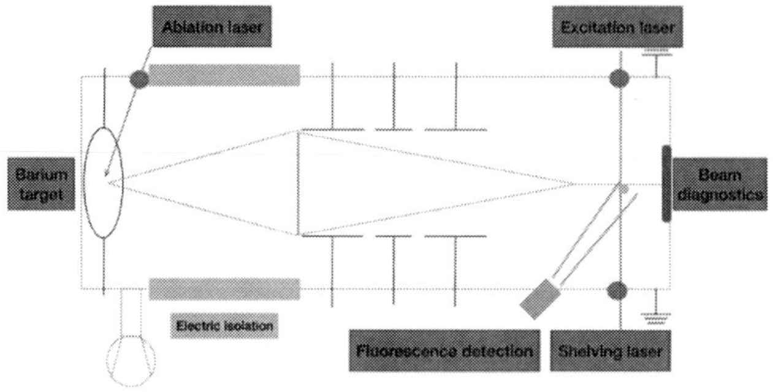

Figure 3. Sketch of the barium ion source vacuum system, consisting of a laser ablation source, an electrostatic isolation, an electrostatic einzellens, 2 lasers for polarization, and a multichannel plate for beam diagnostics.

feasibility of such a storage ring, we plan on storing barium ions at first, which have no nuclear spin and only one valence electron and can be used to test axion-electron couplings. We will be able to set new laboratory limits on axion-electron couplings at the first "switch-on".

(iii) *For polarization* we plan to use a 650 nm laser and a 493 nm laser due to the metastable state [33]. The Zeeman splitting of the ground state in earth's magnetic field is 1 MHz, large enough to polarize the ions without further magnetic fields [34]. We plan on polarizing the ions directly in the ring for the final setup. The feasibility of optical pumping in a storage ring was already observed at the GSI [35], where the observations were explained with rate equations. Adapting these to barium ions, we find that the particles are polarized after about 8 s, which fits well within our demonstrator setup, where we can store the ions for up to 1h.

(iv) *Long spin coherence times* for the ions in the storage ring are required to perform spin precession measurements. We used a ballistic estimation for the T_2 time from polarized ions in a box [36].

$$\frac{1}{T_2} \sim \frac{4\gamma^2 R^4}{175D}(\nabla B)^2, \tag{1}$$

with an estimation for the diffusion constant in the low-pressure limit as $D = v\frac{R^3}{\lambda} = 4.17 \cdot 10^{-3}\,\frac{m^2}{s}$. We used the T_2 time estimations for ions stored in a box here as an approximation assuming particles in a bunch behave similarly. We leave a detailed estimation for later. Estimating all contributions to the magnetic field gradients, we arrive at a T_2 time of about 21 min. This could be further improved by having a more focused beam and a smaller beam divergence, but it does not limit the first measurements with the demonstrator setup.

Table 1. Relevant parameters for the demonstrator storage ring.

Beam energy	β	$\gamma_{Ba+}/2\pi$	Ring frequency	Storage time	Number of ions
30 keV	0.3	$28\,\frac{GHz}{T}$	36 kHz	5 h	10^9

The spin projection noise gives the fundamental limit on the magnetic field measurement sensitivity

$$\delta B \sim \frac{1}{2\mu_{Ba} t\sqrt{N}} \sim 10^{-20}\,T, \tag{2}$$

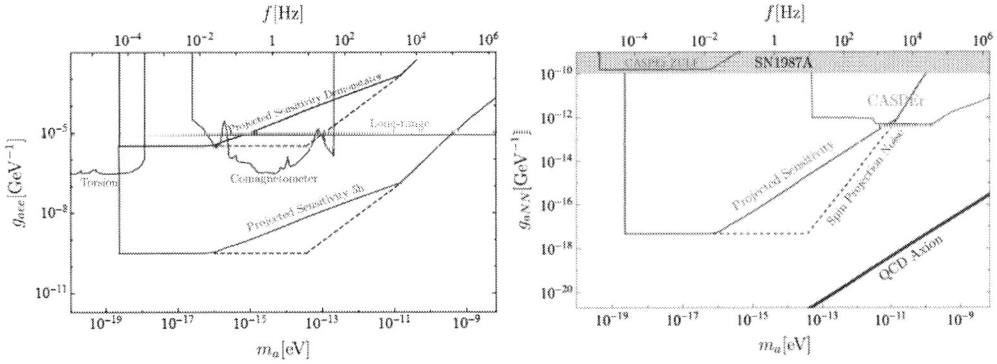

Figure 4. Projected sensitivity of the storage ring. Axion-electron coupling (left), with storage times of 1s (blue) and 5h (red), where we show current bounds from long-range interactions[37], torsion pendulum [38] as well as comagnetometer bounds [39]. Note that the red line approaches the astrophysical bounds [40]. The total scanning time for the blue line is $\sim$ 2h and for the red line $\sim$ 4 years. Axion-nucleon couplings (right) for a final stage with 5h storage time and a nuclear spin polarized xenon sample in the center of the ring. The projections of the CASPEr experiment are taken from [3, 4].

where we use the parameters of Tab. 1.

4 Projected sensitivities

4.1 EDM measurements with a cryogenic ring

A naive rescaling from an experiment trying to measure the electron EDM using molecular ions confined in a radio frequency trap [41] to our setup with 10^9 particles and 5 h storage time leads to an increased sensitivity of 10^6 in statistics. This needs to include all possible and expected new systematic issues emerging from transferring a rather sophisticated experimental technique to a novel and quite a different setting and yet unknown difficulties.

4.2 Axion wind experiment

If the axion contributes to the observed dark matter abundance, its couplings to electron and nuclear spins can be interpreted as an oscillating effective background magnetic field [7, 42],

$$\mathcal{L}_{int} \supset g_{a\psi\psi}\partial_\mu a\bar{\psi}\gamma^5\gamma_\mu\psi \rightarrow H = -g_{a\psi\psi}\nabla a \cdot \sigma. \quad (3)$$

This effective magnetic field induces a spin precession in the storage ring, which can be measured either with SQUIDs or with an optical readout. The demonstrator setup can directly probe the axion-electron coupling (Fig. 4 left panel), while for a final setup, a sample of nuclear spin polarized xenon atoms with density $n \sim 10^{25}\mathrm{m}^{-3}$ and polarization $P \sim 0.3$ can be placed in the center of the ring. The projected sensitivity on the axion-nucleon coupling via a double resonant search is shown in Fig. 4 (right panel). Note that to arrive at the above bound, we needed to chose a particular scanning scheme; we use the full 5 h coherent measurement time for low axion masses, while for the demonstrator we need only 1 s coherent measurement time. As there is no magnetic shielding of the storage ring, the same search probes kinetically mixed dark photon dark matter [43].

5 Conclusion

A concept for an electrostatic table-top-sized storage ring for polarized particles as a new type of magnetometer is discussed. While experimentally challenging, its shot-noise limited sensitivity is 10^{-20} T at < mHz bandwidth, making it extraordinarily sensitive for specific applications.
Among the most exciting use cases is the measurement of effective magnetic fields caused by light-dark matter particles with different types of spin-dependent couplings, which are seen as effective time varying magnetic fields in the experiment. The magnetometer can also be used as a container for a large number of ionic molecules for electric dipole moment measurements, enabling hours of storage and coherence time with the possibility to control and investigate systematic effects.
An experimental realization of the first stage of the experiment, including the particle beam and polarization of trapped ions, is currently starting. The barium ion source will be finished beginning next year, and the storage ring vacuum chamber will be set up next year.

This research was supported by the Excellence Cluster ORIGINS, funded by the Deutsche Forschungsgemeinschaft (DFG, German Research Foundation) under Germany's Excellence Strategy – EXC-2094 – 390783311. The work of SS is additionally supported by the Swiss National Science Foundation under contract 200020-18867.

References

[1] P.W. Graham and S. Rajendran, Phys. Rev. D **88**, 035023 (2013)
[2] D. Budker et al., Phys. Rev. X **4**, 021030 (2014)
[3] A. Garcon et al., Sci. Adv. **5**, eaax4539 (2019)
[4] J. Kimball et al., Springer Proc. Phys. **245**, 105–121 (2020)
[5] S.P. Chang et al., Phys. Rev. D **99**, 083002 (2019)
[6] O. Kim et al., Phys. Rev. D **104**, 096006 (2021)
[7] P. W. Graham et al., Phys. Rev. D **103**, 055010 (2021)
[8] P. Agrawal et al., arXiv: 2210.17547
[9] F. Abusaif et al., CERN Yellow Rep.Monogr. **3** (2021)
[10] A. V. Gramolin et al., Nature Phys. **17**, 1 (2021)
[11] A. Arvanitaki et al., arXiv: 2112.11466
[12] C. Gao et al., arXiv: 2208.14454
[13] M. Lisanti et al., Phys. Rev. D **104**, 055037 (2021)
[14] P. Sikivie, Phys. Rev. Lett. D **113**, 201301 (2014)
[15] M. Jiang et al., Nature Phys. **17**, 1402–1407 (2021)
[16] P. W. Graham et al., Phys. Rev. D **97**, 075020 (2018)
[17] K. Gaul et al., Phys. Rev. Lett. **125**, 123004 (2020)
[18] A. Berlin and K. Zhou, arXiv: 2209.12901
[19] C. Abel et al., Phys. Rev. X **7**, 041034 (2017)
[20] J. Pretz et al., EPJ C **80**, 107 (2020)
[21] A.J. Silenko, EPJ C **82**, 856 (2022)
[22] H. Davoudiasl and R. Szafron, arXiv: 2210.14959
[23] T.S. Roussy et al., Phys. Rev. Lett. **126**, 171301 (2021)
[24] Y. V. Stadnik, arXiv: 2206.10808
[25] C. Smorra et al., Nature **575**, 310–314 (2019)

[26] T. E. Chupp et al., Rev. Mod. Phys. **91**, 015001 (2019)
[27] K. Schindl, *Space charge* (2006)
[28] K. Blaum, MPI Heidelberg, *private communications*
[29] R. von Hahn et al., Rev. Sci. Instrum. **87**, 063115 (2016)
[30] G. Guidoboni et al. (JEDI Collaboration), Phys. Rev. Lett. **117**, 054801 (2016)
[31] H.Wollnik et al., AIP Conf. Proc. **177**, 74-85 (1988)
[32] K. Murray et al., IJMS **472**, 116763 (2022)
[33] B. Mong et al., Phys. Rev. A **91**, 022505 (2015)
[34] C. M. Seck et al., Phys. Rev. A **93**, 5 (2016)
[35] W. Nörtershäuser et al., Phys. Rev. Accel. Beams **24**, 024701 (2021)
[36] G.D. Cates et al, Phys. Rev. A **37**, 2877–2885 (1988)
[37] W. A. Terrano et al., Phys. Rev. Lett. **115**, 201801 (2015)
[38] W. A. Terrano et al., Phys. Rev. Lett. **122**, 231301 (2019)
[39] I.M. Bloch et al, JHEP **2020**, 167 (2020)
[40] M. Bertolami et al., JCAP **10**, 069 (2014)
[41] W. B. Cairncross et al., Phys. Rev. Lett. **119**, 153001 (2017)
[42] Y. V. Stadnik and V. V. Flambaum, Phys. Rev. D **89**, 043522 (2014)
[43] A. Nelson and J. Scholtz, Phys. Rev. D **84**, 103501 (2011)

ENUBET: a monitored neutrino beam for high precision cross section measurements

Marta Torti[3,*], *Fabio* Acerbi[1], *Ioannis* Angelis[21], *Luca* Bomben[2,3], *Maurizio* Bonesini[3], *Filippo* Bramati[3,4], *Antonio* Branca[3,4], *Claudia* Brizzolari[3,4], *Giulia* Brunetti[3,4], *Marco* Calviani[6], *Stefano* Capelli[2,3], *Sara* Carturan[7], *Maria Gabriella* Catanesi[8], *Simone* Cecchini[9], *Nikolaos* Charitonidis[6], *Federico* Cindolo[9], *Giacomo* Cogo[10], *Gianmaria* Collazuol[5,10], *Flavio* Dal Corso[5], *Claudia* Delogu[5,10], *Gianfranca* De Rosa[11], *Andrea* Falcone[3,4], *Brennan* Goddard[6], *Alberto* Gola[1], *Leon* Halić[20], *Fabio* Iacob[5,10], *Cecile* Jollet[12,14], *Verena* Kain[6], *Budimir* Klicek[20], *Yury* Kudenko[13], *Christos* Lampoudis[21], *Marco* Laveder[5,10], *Andrea* Longhin[5,10], *Lucio* Ludovici[15], *Evgenii* Lutsenko[2,3], *Lorenzo* Magaletti[8], *Gianni* Mandrioli[9], *Anselmo* Margotti[9], *Valerio* Mascagna[22,23], *Nicoletta* Mauri[9], *Luca* Meazza[3,4], *Anselmo* Meregaglia[14], *Mauro* Mezzetto[5], *Marzio* Nessi[6], *Alessandro* Paoloni[17], *Michelangelo* Pari[5,10], *Elisabetta Giulia* Parozzi[3,4], *Laura* Pasqualini[9,18], *Giovanni* Paternoster[1], *Laura* Patrizii[9], *Michele* Pozzato[9], *Michela* Prest[2,3], *Fabio* Pupilli[5], *Emilio* Radicioni[8], *Alan Cosimo* Ruggeri[11], *Dimos* Sampsonidis[21], *Carlo* Scian[10], *Gabriele* Sirri[9], *Mario* Stipcevic[20], *Matteo* Tenti[9], *Francesco* Terranova[3,4], *Spyros* Tzamarias[21], *Erik* Vallazza[3], *Francesco* Velotti[6], and *Lucia* Votano[17]

[1]Fondazione Bruno Kessler (FBK) and INFN TIFPA, Trento, Italy
[2]DiSAT, Università degli studi dell'Insubria, Como, Italy
[3]INFN, Sezione di Milano-Bicocca, Milano, Italy
[4]Phys. Dep. Università di Milano-Bicocca, Milano, Italy
[5]INFN Sezione di Padova, Padova, Italy
[6]CERN, Geneva, Switzerland
[7]INFN Laboratori Nazionali di Legnaro, Legnaro (PD), Italy
[8]INFN Sezione di Bari, Bari, Italy
[9]INFN, Sezione di Bologna, Bologna, Italy
[10]Phys. Dep. Università di Padova, Padova, Italy
[11]INFN, Sezione di Napoli, Napoli, Italy
[12]IPHC, Université de Strasbourg, CNRS/IN2P3, Strasbourg, France
[13]Institute of Nuclear Research of the Russian Academy of Science, Moscow, Russia
[14]Centre de Etudes Nucleaires de Bordeaux Gradignan, Bordeaux, France
[15]INFN, Sezione di Roma 1, Rome, Italy
[16]CENBG, Université de Bordeaux, CNRS/IN2P3, Gradignan, France
[17]INFN, Laboratori Nazionali di Frascati, Frascati (Roma), Italy
[18]Phys. Dep. Università di Bologna, Bologna, Italy
[19]Phys. Dep. Università degli Studi di Napoli Federico II, Napoli, Italy
[20]Center of Excellence for Advanced Materials and Sensing Devices, R. Boskovic Institute, Zagreb, Croatia
[21]Aristotle University of Thessaloniki, Thessaloniki , Greece
[22]DII, Università degli studi di Brescia, Brescia, Italy
[23]INFN, Sezione di Pavia, Pavia, Italy

*email address: marta.torti@cern.ch

Abstract. The main source of systematic uncertainty on neutrino cross section measurements at the GeV scale is represented by the poor knowledge of the initial flux. The goal of cutting down this uncertainty to 1% can be achieved through the monitoring of charged leptons produced in association with neutrinos, by properly instrumenting the decay region of a conventional narrow-band neutrino beam. Large angle muons and positrons from kaons are measured by a sampling calorimeter on the decay tunnel walls (tagger), while muon stations after the hadron dump can be used to monitor the neutrino component from pion decays. This instrumentation can provide a full control on both the muon and electron neutrino fluxes at all energies. Furthermore, the narrow momentum width (<10%) of the beam provides a O(10%) measurement of the neutrino energy on an event by event basis, thanks to its correlation with the radial position of the interaction at the neutrino detector. The ENUBET project has been funded by the ERC in 2016 to prove the feasibility of such a monitored neutrino beam and is cast in the framework of the CERN neutrino platform (NP06) and the Physics Beyond Colliders initiative. In our contribution, we summarize the ENUBET design, physics performance and opportunities for its implementation in a timescale comparable with next long baseline neutrino experiments.

1 The ENUBET monitored neutrino beam

The main source of systematic uncertainty on the neutrino cross-section comes from the knowledge of the initial flux in conventional neutrino beams. The ENUBET idea is to develop the concept of monitored neutrino beams [1], where a controlled neutrino flux can be obtained by measuring the charged leptons from mesons along the decay tunnel.
The physics goal of ENUBET is to use the monitoring technique to reduce the uncertainty on the flux of ν_e and ν_μ below 1%, allowing a high-precision determination of neutrino cross sections at the energy of interest for DUNE and HyperK. It will, therefore, reduce the systematics of long-baseline experiments increasing the sensitivity to oscillation parameters and to CP violating phase.
ENUBET started as an ERC project in June 2016 with with the purpose of developing a facility for the measurement of large angle positrons from the $K^+ \rightarrow e^+ \pi^0 \nu_e$ decay, which is directly linked to the flux of ν_e. Furthermore, in March 2019, it was approved as a CERN Neutrino Platform experiment (NP06), in which it also considers the possibility of measuring muons both from the two body large angle decay of the kaons ($K^+ \rightarrow \mu^+ \nu_\mu$) and, after the hadron dump, from the pion decays ($\pi^+ \rightarrow \mu^+ \nu_\mu$) [2].

2 The ENUBET beamline and performance

ENUBET is a conventional neutrino beam where the pions and kaons are produced by protons impinging on a fixed target. They are focused and momentum selected with a momentum bite of O(5 - 10%), by a short transfer line (TL) of about ~ 20 m. The TL is designed for a 8.5 GeV/c particle beam, an optimal choice for the e^+/π^+ separation and to span the region of interest for future neutrino oscillation experiments (DUNE and HyperK). A complete G4beamline[3] simulation is performed and permits to evaluate the beam composition, while the shielding of the components has been optimized with FLUKA and GEANT4. The latest versions of the TL implement a double dipole configuration, that allows to reduce backgrounds from stray particles and the ν_e component at the neutrino detector from early decays in the first part of the TL. The selected particles are then transported to the decay tunnel that is located off the axis of the proton beam. Non-interacting protons are stopped in a proton dump. The 40 m long decay tunnel is instrumented along its walls to monitor the leptons. Not decayed particles (mostly pions) and leptons produced along the axis (mostly muons from pion decay) are stopped by a hadron dump at the end of the tunnel.

Assuming $\sim 10^{20}$ proton on target, ENUBET is able to provide a sample of about 10^4 ν_e CC (Figure 1 left) and 10^6 ν_μ CC in a 500 t detector located 50 m after the hadron dump monitored with an expected precision of 1%.

The other source of systematic uncertainty is the reconstruction of the ν energy, which is biased by the inaccurate reconstruction of the final state particles. ENUBET is a narrow band beam and using the off-axis narrow band technique [4]. It can provide a measurement of the neutrino energy at 10-20% level, just by locating the radial position of the interaction vertex and without relying on accurate final state particle reconstruction (Figure 1 right).

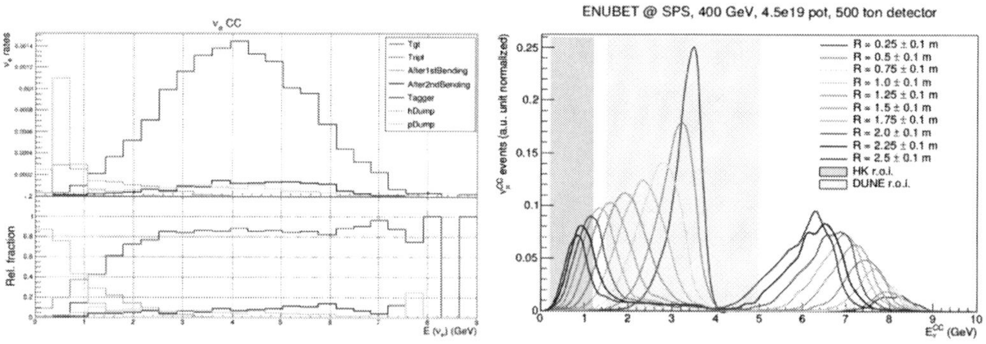

Figure 1. Left top: energy spectrum for ν_e CC interactions in a detector of 500 ton mass and 6× 6 m^2 transverse size placed 50 m from the decay tunnel end. Each color corresponds to a neutrino contribution produced within a region of the beamline. Left bottom: relative fraction of each category with respect to the total ν_e CC rate. Right: ν_μ CC energy spectra for different values of distance (R) between the neutrino interaction vertex and the beam axis .

3 Tunnel instrumentation

The ENUBET tunnel instrumentation (see Figure 2) is based on a calorimeter for e^+ / π^+ separation and on an inner light-weight photon veto for e^+ / π^0 separation. This last detector also provides the absolute timing of the events and is thus called the "t_0-layer". Between 2016 and 2018 several prototypes have been built and tested at the CERN East Experimental Area [5–8] to validate the physics performance.

The final choice is an iron-scintillator sampling calorimeter divided into modules (Lateral readout Compact Module, LCM) of 10 cm length that samples the showers every 4.3 radiation lengths. The light is collected by WLS fibers running along the lateral edges of the five scintillator tiles of each module and bundled in groups of ten, and read out by SiPMs.

The photon veto is based on plastic scintillators, whose light is transported by WLS fibers toward the upper part, beyond a 40 cm thick borated polyethylene shielding, where all the SiPMs are located. A detailed description of the performance of a small scale prototype based on this concept is given in [9].

In order to prove the effectiveness of the technique and validate the detector performance, a large scale prototype (the ENUBET "demonstrator" [10]) of 1.65 m length and 3.5 ton mass has been built. It was exposed to the T9 particle beam at CERN PS in Fall 2022.

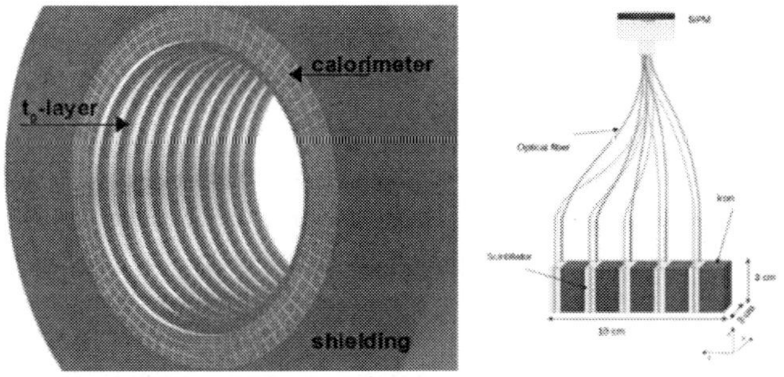

Figure 2. Left: scheme of the ENUBET decay tunnel, instrumented with three layers of calorimetric modules (green), an inner t_0-layer of scintillator tile doublets (yellow) and an outer shielding of Borated Polyethylene (brown). Right: layout of a single Lateral readout Compact Module (LCM).

4 Particle identification and assessment of systematics

A full GEANT4 simulation of the ENUBET facility has been implemented and it includes particle propagation and decay from the transfer line to the detector. The simulation has been validated with data from the prototype tests.

Positrons and muons in the calorimeter are identified with dedicated algorithms [11]. The event reconstruction starts with the identification of a seed, that corresponds to a hit in a LCM of the innermost layer. All LCM and t_0-layer energy deposits compatible in space and time with the seed are clustered together and constitute the candidate event. A set of variables describing the energy deposition in the calorimeter and the event topology are used as input for a Neural Network, based on the Root TMVA package [12], that identifies and classifies positrons and muons.

The preliminary performance (Figure 3) obtained for e^+ from K_{e3} show a selection efficiency of about 22% dominated by the geometrical efficiency ($\sim$ 50%), and a S/N of ~2.1, while μ from K^+ are reconstructed with an efficiency of about 34%, and a S/N of ~6.0 .

The relation between leptons measured in the calorimeter and neutrino flux at the detector is used to evaluate neutrino flux systematics. A signal plus background model is built, including a priori hadroproduction and TL related systematic uncertainties. Toy-MC experiments are produced and fitted to study the a posteriori systematic uncertainty and assess the corresponding uncertainty reduction.

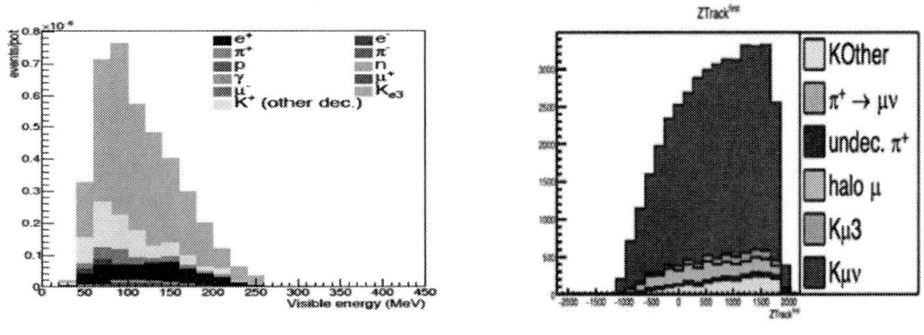

Figure 3. Reconstructed visible energy of K_{e3} positron events (left) and reconstructed impact point along decay tunnel calorimeter wall (Z) of $K_{\mu\nu}$ (right) after the application of the Neural Network.

5 Conclusions

The ENUBET main idea is to prove the feasibility of a monitored neutrino beam, where charged leptons, produced in association with neutrinos, are monitored in order to estimate the neutrino flux. This goal can be achieved by properly instrumenting the decay region of a conventional narrow-band neutrino beam, using a sampling calorimeter. The ENUBET Collaboration proved the feasibility and cost-effectiveness of this technology whose complexity does not exceed significantly the one of a conventional short baseline beam. It was also devised the first end-to-end simulation of a monitored beam and a full Geant4 simulation of the facility is employed to assess the final systematics budget on the neutrino fluxes. The ENUBET results will play an important role in the systematic reduction programme of future long baseline experiments, thus enhancing the physics reach of DUNE and HyperKamiokande.

References

[1] A. Longhin, L. Ludovici and F. Terranova, Eur. Phys. J. **C 75** 155 (2015); A. Berra et al., CERN-SPSC-2016-036, SPSC-EOI-014, Geneva (2016).
[2] F. Acerbi et al., CERN-SPSC-2020-009, SPSC-SR-268, Geneva (2020).
[3] T. J. Roberts et al., Proc. EPAC'08 2776-2779 (2008).
[4] F. Acerbi et al., CERN-SPSC-2018-034, SPSC-I-248, Geneva (2018).
[5] A. Berra et al., Nucl. Instrum. Meth. **A 830** 345 (2016).
[6] A. Berra et al., IEEE Trans. Nucl. Sci. **64** 1056 (2017).
[7] G. Ballerini et al., JINST **13** P01028 (2018).
[8] F. Acerbi et al., JINST **14** P02029 (2019).
[9] F. Acerbi, et al. JINST **15** P08001 (2020).
[10] F. Acerbi et al., CERN-SPSC-2021-013, Geneva (2021).
[11] F. Pupilli et al., PoS NuFact2021 027 (2022).
[12] A. Hoecker et al., *TMVA-Toolkit for Multivariate Data Analysis* arXiv:physics/0703039

AUTHOR INDEX

AUTHOR INDEX